P9-ECJ-616

# DIGITAL HUSTLERS

# DIGITAL HUSTLERS

## LIVING LARGE AND FALLING HARD IN SILICON ALLEY

### CASEY KAIT AND STEPHEN WEISS

ReganBooks
*An Imprint of* HarperCollins*Publishers*

# FOR OUR PARENTS

www.digitalhustlers.com

DIGITAL HUSTLERS. Copyright © 2001 by Casey Kait and Stephen Weiss. All rights reserved. Printed in the United States of America. No part of this book may be used or reproduced in any manner whatsoever without written permission except in the case of brief quotations embodied in critical articles and reviews. For information address HarperCollins Publishers Inc., 10 East 53rd Street, New York, NY 10022.

HarperCollins books may be purchased for educational, business, or sales promotional use. For information please write: Special Markets Department, HarperCollins Publishers Inc., 10 East 53rd Street, New York, NY 10022.

FIRST EDITION

Printed on acid-free paper

Library of Congress Cataloging-in-Publication Data

Kait, Casey.
    Digital hustlers : living large and falling hard in Silicon Alley / Casey Kait and Stephen Weiss.— 1st ed.
        p.   cm.
    ISBN 0-06-620923-4
    1. Internet industry—New York State—New York—Case studies.   2. Entrepreneurship—New York State—New York—Case studies.   I. Weiss, Stephen.   II. Title.
HD9696.8.U63 N75 2001
338.4'61004678'097471—dc21

                                                                        2001019192

01   02   03   04   05   RRD   10   9   8   7   6   5   4   3   2   1

# CONTENTS

# INTRODUCTION

In the early 1990s New York City was stuck in a recession, still struggling to recover from the hangover of the Reagan years. The Wall Street wilding of the mid '80s seemed a distant and distasteful memory, a time when new money heroes were made and discarded through junk bonds and insider trading.

Though the recent college graduates of the early '90s had the benefit of being well read and well rounded, their options were limited. The creative jobs they longed for were few and far between, and many found the notion of becoming a drone for a faceless corporation too dispiriting to consider.

Then something happened in California.

Thirty-five miles south of San Francisco, entrepreneur Jim Clark hired twenty-two-year-old Marc Andreessen and a handful of his colleagues from the University of Illinois—most of the original team that, while still students, had built the first successful and easy-to-use Web browser, Mosaic. With Clark's money and some outside capital they started Mosaic Communications—which very quickly became Netscape, the first real Internet company, the company that built the browser that forced Microsoft, America Online, and eventually most of the world onto the Web.

In 1993 the Internet was being used by roughly one hundred thousand people in America to share research information, exchange email, and communicate over bulletin board systems. In New York City, the first generation of young adults who had grown up with personal computers, video games, and cable television—many of them college graduates recently arrived in the city—became fascinated by this new medium. And with a sense of purpose, destiny, and entitlement that defied their slacker image, they jumped into this new world as multimedia artists, online publishers, software developers, and ultimately entrepreneurs.

By 1995, this generation of young New Yorkers had adopted the name Silicon Alley for their loose agglomeration of talent, and the pace of their burgeoning new industry fed off the vigor of the city itself. The rising stars of the Alley tapped into the pulse of the city, capturing the attention of the national media, becoming the toast of Wall Street, and pouring millions of dollars into Madison Avenue's advertising coffers. By the year 2000, Silicon Alley was employing 250,000 workers and producing nearly seventeen billion dollars in revenue. Digital sweatshops proliferated, housing tribes of talented young people at a time of extraordinary opportunity, wild excess, and massive shifts in culture, all fueled by the rapid changes in technology.

The Internet explosion also sired a blur of lavish parties, politicking, and frivolity, as freshly minted millionaires enjoyed the spoils of their success. Clever young company founders took meetings at TriBeCa lunch spots with the long-ball hitters of the New York finance community, bought multimillion-dollar real estate and expensive vintage cars. Geek chic quickly became the new gold standard.

The entrepreneurs of Silicon Alley were a different breed from their Valley counterparts. Though some came from technology backgrounds, many more leaped into the Internet from the arts and from other businesses. Far away from the great technology center of the West, the entrepreneurs of Silicon Alley had to rely on what they knew.

Lured by the excitement, the money, and the creativity, next-generation content producers—writers, artists, designers—began to put their work online. Innovative content sites like Word and FEED (feed mag.com) were among the first kids on the block. The pluckiest of the online magazines took on work for hire to cover their expenses, building websites for corporations making the move online. Money began to pour into fledgling interactive agencies like Razorfish and Agency.com, born alongside the early content sites The Blue Dot and UrbanDesires. Many such content sites were quickly dropped in favor of the lucrative contracts that could be won from Fortune 500 companies.

For everyone involved, it was a leap into the unknown. Still in its infancy, the industry had yet to prove itself as a viable form of business. But the early entrepreneurs were smart, scrappy, and hungry. With accessible new technologies at their disposal, they realized for the first time that they could make a living—even become financially successful—doing something they loved. They were inventing the rules as they went along, and it worked. For a time, anyone who could register a domain name could start a business. And the "sizzle" was selling.

Sharing space with giants in the media center of the world in New York City, the new entrepreneurial class was vaulted into recognition in the national press. When a pair of former lovers, Rufus Griscom and Genevieve Field, turned an idea for an erotic e-zine into a mini media empire with their website Nerve.com, they generated a knee-high stack of newspaper articles, magazine shoots, and television clips that promoted them as the Scott and Zelda of the Internet and spread their story around the world.

Already a multimillionaire from his role as founder of Jupiter Communications, Josh Harris took an even more high-profile approach to Internet stardom. By founding Pseudo.com, the first Webcasting company in the world, he turned his corner of the Internet into a digital playhouse with offbeat and outrageous post-television programming. With the inspiration to record the entire world in digital video, he threw lavish events for artists and cyberpunks, and preserved them on camera.

With a median age under thirty, the denizens of this new industry also displayed an almost compulsive desire to party and network. Even the shyest entrepreneurs were transformed into social butterflies when it came to pitching business at a cocktail party. Slipping business cards into the hands of new friends, would-be investors—even potential lovers—became standard operating procedure.

At his fin de siècle New Year's Eve party, Josh Harris spent over a million dollars, invited artists from around the world and half of the city of New York to visit, and kept the booze flowing and digital video cameras rolling for a full month. Filling six floors of two adjacent downtown warehouses, his revelers formed a bacchanalian society of Gatsbyesque proportions. And as the clock struck midnight in the eighty private bed chambers he had built, couples rang in the new century under the watchful eyes of digital surveillance cameras.

For Internet entrepreneurs, the whole year had been a party. In late December at the end of the twentieth century, things were better than they had ever been. Fast money and high hopes abounded. Every night of the week, New York hot spots like the Roxy played host to throngs of up-and-comers in the industry. At one Willy Wonka–themed fête thrown by DoubleClick, the Alley's wealthiest company, overflowing containers of candy lined the dance floor, Oompa-Loompa busboys hustled blinking trays of drinks, and cigarette girls in vinyl miniskirts wove through the crowd. Excess was the rule; one lavish party followed the last. The digerati kissed each other on the cheeks, exchanging ideas, business cards, and email addresses. Men and women carried their business deals into the bathroom on cell phones, while Internet "business" events seemed to revolve around a constant theme of ice sculptures and sushi.

Like the Internet itself, the culture of Silicon Alley grew so rapidly that it quickly generated its own unique personalities: people like Courtney Pulitzer, a grand-niece of Joseph, whose monthly cocktail parties in opulent settings brought together the leaders of the industry, and Jason McCabe Calacanis, who with his industry print magazine *Silicon Alley Reporter* served as a chronicler and interpreter of the New York new media scene. What started as a black-and-white newsletter in 1996 blossomed within a few short years into an empire of national business conferences, two monthly magazines, five daily and weekly email reports, and the closely watched Silicon Alley 100 list, which ranked the top entrepreneurs and became the most potent status symbol of the Internet elite.

They were the first generation of new media entrepreneurs, digital hustlers unburdened by reverence for the corporate world. The Internet was a truly adaptive medium; digital technology allowed you to capture anything, duplicate it, reuse it in other media, or take it as your own. And it wasn't hard to dream up ways to use that technology and put it toward an age-old occupation: the art of making money in New York.

By 1998 most of the game had already been played, but nobody could stop the momentum. The deals got bigger and bigger, though the returns dwindled.

In March 2000 an article in *Barron's* magazine issued a forecast that over two hundred Internet companies would run out of money in less than two years—and the bubble officially burst. On April 10, a warm Monday in early spring, the NASDAQ dropped nearly three hundred points. Each day of that week, it dropped another one to three hundred points. For the next six months it moved up and down, but almost always more down than up, and by the end of 2000 it had fallen to 2,300, having lost more than half its value.

The Gold Rush was over. Alley companies canceled their IPOs, laid off their employees; many closed their doors altogether. Pseudo.com was just one of many to declare bankruptcy. What had once seemed an environment of endless opportunity was now hostile to growth of any kind. At companies like Razorfish, Agency, iVillage, TheGlobe, Star-Media, and several others, total stock value dropped 95 percent or more. The press, which had once elevated them to celebrity status, now lambasted and mocked the Alley entrepreneurs. Paper fortunes disappeared. The party came to an abrupt halt.

Walking through the hallways of Silicon Alley in the fall and winter of 2000, there was no hiding the gloom that hung over almost every office. There were plenty of empty desks; many of the companies that survived found that their furniture was better than their future. The NASDAQ crash had decimated tech stocks across the country, but with its emphasis on content and client services, Silicon Alley was hit harder than any other digital center in the country. The morale and net worth of Alley employees and founders dropped hand in hand. At companies like Pseudo that went belly-up before ever going public, company shares that had once seemed priceless became worthless overnight. The psychological damage was devastating.

Company officers did their best to seem upbeat, but in reality no one was safe. Every other day the *Silicon Alley Daily* brought news of another company laying off a quarter of its staff or going out of business. The faucets of investment money, which once had seemed to flood the streets of the Alley, had run bone-dry. For companies that had received their last round of funding before April 2000, the clock was ticking: nine months, six months, three months before they would run out of cash. For these private companies—of which there were more than eight thousand at the end of the century—the recipe for success had called for new rounds of funding every three to six months. But by the time they realized they wouldn't be able to raise additional capital, most of their funding had already been spent on the billboards and print ads their investors had demanded.

Today, just a year after the fall of the NASDAQ, the work of many of these companies has all but disappeared. A paper magazine that publishes even once has the potential at least to last forever, tucked away in some attic or shelf. But Web pages are ephemera, existing only when the server is up and running and all the links are operational. When the Silicon Alley servers went dark, the companies quickly faded, and so too did all of the creative content they had once housed.

The Internet had opened a window of opportunity that was never big enough for all the visionaries who tried to climb through it. Some succeeded wildly, beyond any expectation. Others failed so publicly, so ostentatiously, that anyone watching could only marvel as they plummeted back to earth.

Through it all, they experienced more in five years than many are able to in a lifetime. They had it all, and they lost most of it. But they were the first, and for a brief moment they thrived.

# CAST OF
# CHARACTERS

**BERKOWITZ, SUSAN.** Former vice-president, TheGlobe.com. Former senior vice-president of marketing, Wit Capital.

**BORTHWICK, JOHN.** Founder, äda 'web and TotalNY.com. Borthwick's WP Studio delivered early on two successful projects. äda 'web was a digital arts collective and online showcase for Web-based art projects [still online at *www.adaweb.com* through the Walker Art Center in Minneapolis]. TotalNY was one of the first online city guides. Both properties were acquired with WP by AOL's Digital City in February of 1997, reportedly for $5 million in stock. Borthwick is now vice president of new product development at AOL.

**BOWE, MARISA.** Former editor in chief, Word.com. Word was an early online magazine with a focus on fusing stories with groundbreaking Web design. The magazine was started by Icon CMT, but dropped in 1998. Several months later, Zapata, a fish-oil company looking to reinvent itself online with ZAP.com, purchased the rights to Word.com and rehired Bowe and other members of the staff to relaunch the magazine, only to pull the plug in the Summer of 2000. Bowe coedited *Gig: Americans Talk About Their Jobs at the Turn of the Millennium* (Crown, 2000).

**BURNS, RED.** Founder and chair, Interactive Telecommunications Program, New York University. The ITP program, started in 1979, teaches graduate-level classes in interactive media, and has trained several Alley entrepreneurs, including Stacy Horn of Echo.

**BUTTERWORTH, NICHOLAS.** CEO, MTVi group. Former president, SonicNet. Butterworth graduated from Brown in the class of '89, a friend of Steven Johnson and Rufus Griscom. He had already run MTV's Rock the Vote when Tim Nye hired him in 1994 to help build SonicNet into a prominent alternative music BBS in lower Manhattan. Under Butterworth's leadership as president, SonicNet changed hands several times, and reinvented itself as a website for music fans. When Viacom acquired TCI music's interactive assets in 1998, SonicNet passed into its online holdings. Butterworth soon took over the reins of the entire MTVi group, including MTV.com, VH1.com, NICK.com, SonicNet.com, and several international sites. An attempt to take MTVi public in 2000 did not succeed.

**CALACANIS, JASON MCCABE.** CEO and founder, *Silicon Alley Reporter, Digital Coast Reporter,* Rising Tide Studios. In 1996, Calacanis was twenty-five and covering the "digerati" as a social columnist for *Paper* when he printed the first issue of *Silicon Alley Reporter*. Started as a photocopied newsletter, *SAR* quickly expanded into a mini–media empire. Calacanis was a scrappy Brooklyn guy with little journalism experience who saw a niche and jumped at it: in the words of *New Yorker* writer Larissa MacFarquhar, "Calacanis descended upon Silicon Alley in 1996 like Warhol upon a Brillo factory."

**CHERVOKAS, JASON.** Cofounder, @NY. Chervokas was a local reporter in the Bronx in 1995 when he and partner Tom Watson launched @NY as an email-based newsletter covering Silicon Alley. They sold the company to Alan Meckler's Internet.com in 1999, and Chervokas left @NY in late 2000 to join Primedia Ventures as a venture capitalist.

**COLONNA, JERRY.** Cofounder, Flatiron Partners. A former magazine editor, Colonna teamed up with Fred Wilson to form the first major

VC fund focused specifically on Silicon Alley investments. Invested early in StarMedia and several other Alley start-ups.

**CONNORS, CONNIE.** CEO, Connors Communication. The head of one of the leading PR firms in the Alley and an entrepreneur since 1985, Connors was also the president of NYNMA for several years.

**DEROSE, GENE.** Former CEO and chairman, Jupiter Communications, now vice-Chairman and president of Jupiter Media Metrix. DeRose, a former 'zine writer, took over Jupiter from Josh Harris in 1994 and IPO'd the company in October 1999. In the summer of 2000, he merged the online research company with Alley neighbor Media Metrix, which documents Web traffic to sites.

**DUNCAN, THERESA.** CEO, Valentine Media. Duncan has developed content for and produced CD-ROMs, video games, interactive films, and animation.

**DYSON, ESTHER.** Chairman, EDventure Holdings. One of the best-known figures of the Alley, Dyson publishes the influential veteran newsletter *Release 1.0* and wrote a book about living in the digital age called *Release 2.0.* She is a powerful investor in the Alley, and recently ran ICANN, the governing body that administers domain names, as its founding chair.

**EFFRON, JON.** Former sales associate, eGroups.com.

**ESPUELAS, FERNANDO.** CEO and Chairman, StarMedia. StarMedia is one of the leading Spanish-and-Portuguese-language portals in Latin America. Espuelas founded the company in 1996, raised $80 million from Flatiron and Chase two years later, and followed with a $105 million IPO. After the April 2000 shakeout, the stock price fell nearly 90 percent.

**GALINSKY, ROBERT.** Former executive producer, Pseudo Network.

**GOLDSTEIN, SETH.** Cofounder, SiteSpecific. Venture partner, Flatiron Partners. Goldstein was a serial entrepreneur who left a job at Condé Nast to start SiteSpecific.

**GOULD, GORDON.** Former president, *Silicon Alley Reporter* and Rising Tide Studios. Founder and CEO, Upoc.com.

**GRISCOM, RUFUS.** CEO and copublisher, Nerve.com and *Nerve.* Nerve.com is the site for "literate smut," providing erotic stories and images that appeal to both men and women. Griscom founded the company with then-girlfriend Genevieve Field in 1997, shortly before he turned thirty. The two have built Nerve into a vibrant multimedia company, with a print magazine that launched in Spring 2000, several published books, and a thriving personals community.

**HAFT, JEREMY.** Cofounder, SiteSpecific. Now CEO of BchinaB, a company that can be described concisely as "plastics in China."

**HARRIS, JOSH.** Founder, Pseudo.com; founder, Jupiter Communications. Harris founded Jupiter, the online market research giant, out of his Manhattan apartment in 1986. After handing off Jupiter to Gene DeRose in 1994, he started Pseudo, one of the first Internet broadcasting companies, which in its heyday had over fifty shows and two hundred hours of original programming per month, much of it focused on youth and counterculture. The company enjoyed the reputation in the Alley of throwing the wildest parties, but was never able to provide a credible business model. In September of 2000, with the IPO market closed and more than $30 million in investment capital spent, the company laid off its remaining workers and declared bankruptcy. Harris's New Year's Eve 2000 party, which ran for a month in a pair of former fabric factories downtown, cemented his contribution to life in the Alley.

**HEIFERMAN, SCOTT.** Founder, i-Traffic and RocketBoard. Heiferman was a recent graduate of the University of Iowa working for Sony when he started i-Traffic in 1995. Though the online media buyer was acquired by Agency.com just before its late '99 IPO, Heiferman's second venture, RocketBoard, failed to live out its first year.

**HIDARY, JACK.** Cofounder and CEO, EarthWeb. Hidary was twenty-seven and a medical researcher at the National Institutes of Health when he caught the Internet bug and started EarthWeb, first a consulting and developing shop and later a network of sites for IT professionals, with his brother Murray and friend Nova Spivak in 1996.

**HOREY, BRIAN.** Cofounder, New York New Media Association. Partner, Lawrence, Smith and Horey. Horey was not only an investor in the Alley, he was also very focused on building the early community.

**HORN, STACY.** Founder, Echo. Horn graduated from the ITP program and started Echo in the spring of 1990 as an intellectual BBS community.

**IRVINE, CELLA.** Chair, NYNMA. Former general manager of Sidewalk.com in New York.

**JOHNSON, RICHARD.** Founder and CEO, HotJobs.com. Johnson "bet the farm" in 1999 when he mortgaged his house to buy a single Super Bowl spot, a move that propelled the company from noncontender status into a successful IPO and a dead heat with Monster.com for top job site.

**JOHNSON, STEVEN.** Cofounder, FEED (www.feedmag.com); author, *Interface Culture* (HarperEdge, 1996). FEED was one of the original online magazines, and Johnson, a comparative-literature student turned online publisher, started the company with Stefanie Syman. In 2000, they rolled the company up with Suck.com and Alt.Culture to start Automatic-Media.

**JOSELEVICH, BERNARDO.** Founder of "Bernardo's List," the weekly party list for Silicon Alley events.

**KANARICK, CRAIG.** Cofounder and chief scientist, Razorfish. One of the first "media darlings" of the Alley, Kanarick made a lasting impression with clients and investors with his multicolored hair, outrageous outfits, and evangelical belief in the power of the website. Razorfish was perhaps the most visible Web development company, but after April of 2000, the stock fell drastically.

**KAWOCHKA, MICHAEL.** Former producer, TheGlobe.com.

**KRIZELMAN, TODD.** Former co-CEO, TheGlobe.com. On the day of its IPO, TheGlobe realized the greatest increase in first-day share price in the history of Wall Street, rising to nearly $100 per share and elevating the twenty-five-year-old Cornell graduate from a net worth of zero to more than $100 million. As the stock began to tumble to a fraction of a dollar, Krizelman and cofounder and co-CEO Stephan Paternot were forced to step down.

**KURNIT, SCOTT.** Founder and CEO, About.com. About.com is the leading "human guide" to the Web, with hundreds of section editors responsible for individual topics. In October of 2000, Kurnit successfully sold the company to Primedia for $700 million, rescuing it from a drop in stock price of over 75 percent from the previous year. Kurnit was an experienced cable executive (Viacom and Time Warner) and early Alley top dog, the executive vice-president at Prodigy and the head of the failed MCI/Newscorp iGuide project.

**LEVITAN, ROBERT.** Cofounder, iVillage.com. Founder and CEO, Flooz.com.

**LEVY, JAIME.** CEO, Electronic Hollywood. Levy parlayed her design and production skills into digital magazines first distributed on floppy disks, and threw early CyberSlacker parties in the East Village that brought the community together.

**LIU, DAVID.** Cofounder and CEO, TheKnot.com. Liu, who had previously started a CD-ROM company with his partners, launched TheKnot, a popular wedding site, in 1996 with seed funding from AOL. The company went public in 1999 and extended its offerings with books, a magazine, and an online bridal registry.

**MECKLER, ALAN.** Founder and CEO, Internet.com. Founder and former CEO, *Internet World*. Meckler, an entrepreneur with a background in library-newsletter publishing, started the magazine *Internet World* in 1992 and built a vibrant business with conferences like Fall Internet World at the Javits Center, which he eventually sold to Penton Media for nearly half a billion dollars. He simultaneously purchased from Penton the Internet assets to his business, called Internet.com, a network of business sites covering the Internet, which he took public in 1999.

**NAPOLI, LISA.** Reporter, MSNBC. After stints at Prodigy and iGuide, Napoli became a leading Internet commentator for MSNBC.

**NYE, TIM.** Founder, SonicNet.com, Sunshine Interactive, and AllTrue.com. A serial entrepreneur with an interest in performance art, Nye started the Thread Waxing Space, a downtown performance space, before turning his interest in multimedia to the Internet with the formation of SonicNet, a dial-up BBS for alternative-music fans. His recent venture, AllTrue.com, screens reality-based video clips produced by users.

**NYHAN, NICK.** Founder, Dynamic Logic. Dynamic Logic is an online research firm, and Nyhan was formerly at Poppe Tyson (Modem Media).

**O'CONNOR, KEVIN.** Founder and chairman, DoubleClick. DoubleClick is the largest online advertising network in the world and with nearly 2,000 employees, the biggest employer in Silicon Alley, serving five billion banner ads per week. Perhaps the most powerful of all the Silicon Alley companies, DoubleClick found itself at the center of a public debate on Internet privacy and security in January of 2000 after it acquired a data mining company to help it track Web surfers on- and offline.

**O'ROURKE, ALICE RODD.** Executive director, NYNMA. A former head of the New York State Department of Economic Development, O'Rourke took over NYNMA in 1998.

**PAGKALINAWAN, CECILIA.** Founder, Boutique Y3K, an e-commerce consulting company for fashion companies. Pagkalinawan bought her company for a dollar when its parent company tried to dump it.

**PATERNOT, STEPHAN.** Former co-CEO, TheGlobe.com.

**PRESTON, JON EGAN.** Former systems administrator and host of ParseTV, Pseudo.

**PULITZER, COURTNEY.** CEO, Courtney Pulitzer Creations, Cocktails with Courtney. Called the "Doyenne of Silicon Alley" and "TriBeCa Contessa," the ever-convivial Pulitzer leveraged her Web design work into a full-time hostessing gig as the queen of the Internet party scene. Her monthly "Cocktails with Courtney" gathering drew scores of people in the industry, her weekly report on the CyberScene has thousands of subscribers, and she has expanded her operations to include cocktail parties in dozens of cities around the world.

**RASIEJ, ANDREW.** Cofounder, Digital Club Network. Founder, MOUSE.org (Making Opportunities for Upgrading Schools and Education). DCN cybercasts live music performances from clubs. MOUSE, a public-service organization that wires public schools to the Internet, has 1,500 volunteers. Rasiej formerly owned Irving Plaza, a concert hall near Union Square.

**RUSHKOFF, DOUGLAS.** Writer, commentator. Rushkoff, the author of *Coercion: Why We Listen to What They Say* and *Ecstasy Club*, is a regular contributor to National Public Radio and one of the more outspoken commentators in Silicon Alley.

**RYAN, KEVIN.** CEO, DoubleClick. Ryan joined DoubleClick as its twentieth employee after running Dilbert.com, but was named CEO of the company in 2000.

**SHANNON, KYLE.** Cofounder and chief people officer, Agency.com; founder, UrbanDesires; founder, WWWAC (World Wide Web Artists

Consortium). A restless mind must have led this former actor to start all three of these projects, many with partner Chan Suh.

**SHIRKY, CLAY.** Former CTO, SiteSpecific. New-media professor, Hunter College. Frequent Internet commentator. Coauthored one of the first books on HTML. His essays about technology are collected at www.shirky.com.

**SILVERMAN, BEN.** Editor, DotComScoop. DotComScoop was started in 2000 as a wireless messaging group on Upoc.com, focused on breaking dot-com stories, most of which, after the Spring 2000 crash, were "bad breaks."

**SINGER, MARC.** Cofounder and former chief creative officer, togglethis. The togglethis take on online marketing was that it should be interactive, and successful campaigns included a Virgin Airlines download with Austin Powers, and a Bozlo Beaver interactive character sold to Warner Bros.

**STAHLMAN, MARK.** Cofounder, NYNMA. Credited with inventing the term "Silicon Alley." In 1992, Stahlman was young, retired, and living in a gorgeous loft in the Flatiron district, a site that would serve as the early home for Cyber Salons and NYNMA board meetings.

**SYMAN, STEFANIE.** Cofounder, FEED. Before starting FEED with Steven Johnson, she was a twenty-five-year-old *Wall Street Journal* reporter.

**TANG, SYL.** President, HipGuide.com. Tang left Andersen Consulting to start her company in 1999.

**WASOW, OMAR.** Executive director, Blackplanet.com. After graduating from Stanford, Wasow started Brooklyn BBS NewYorkOnline and later became an Internet commentator for MSNBC. Wasow was tapped in 1999 to run Blackplanet, a popular community site based in SoHo. He catapulted to fame as "Oprah's computer guy" when he helped her go online for the first time as part of a multipart series. Wasow is one of the most popular figures in the Alley.

**WHEATLEY, ANNA.** Cofounder and editor in chief, *AlleyCat News*. With partner Janet Stites, Wheatley founded the magazine with a focus on investment opportunities and financial analysis of Alley companies.

**WILSON, FRED.** Cofounder, Flatiron Partners.

**YOUNG, JOHN.** Former president, Tribal DDB. Young also worked at Poppe Tyson (Modem Media).

**ZAINO, JESS.** Former host of the celebrity show *StarFreaky* on Pseudo.com. Zaino had not graduated college when she found Pseudo, and a dream job, in 1998.

**ADDITIONAL PEOPLE MENTIONED IN TEXT:**

**ABRAHAMSON, KURT.** President and COO, Jupiter Media Metrix.

**ADAMO, DENNIS.** Former executive producer, Pseudo.

**ANDIORIO, ANNE.** Senior vice-president, StarMedia Network.

**ANDREESSEN, MARC.** Mosaic builder and Netscape cofounder.

**ASNES, TONY.** Former president, Pseudo.com.

**BAXTER, SCOTT.** Founder and CEO, Icon CMT. Founder and CEO, Hawk Holdings.

**BEJAN, ROBERT.** Director of worldwide sales, Microsoft.

**BELL, MARC.** Founder and CEO, Globix, a high-tech hosting and colocation facility based in lower Manhattan.

**BENNETT, ED.** Former CEO, Prodigy.

**BERNERS-LEE, TIM.** Inventor of World Wide Web.

**BEZOS, JEFF.** Founder and CEO, Amazon.com.

**BOHRMAN, DAVID.** Former CEO, Pseudo.com.

**BOHNETT, DAVID.** Founder, GeoCities.

**BUNN, AUSTIN.** Reporter, *Village Voice* and *FEED*.

**DAVID BYMAN.** Former CTO, SiteSpecific.

**CARLICK, DAVE.** Former senior vice president, Poppe Tyson, and cofounder, DoubleClick. Partner, VantagePoint Venture Partners.

**CARPENTER, CANDICE.** Cofounder and Former CEO, iVillage.

**CAVELLA, JOEY.** Creative director, Nerve.com.

**CHEN, JACK.** Cofounder and president, StarMedia.

**CLARK, JIM.** Founder, Silicon Graphics, Netscape, Healtheon.

**COHEN, DILLON.** Investor, the Carlin Group.

**CHUBB, SARAH.** CEO, CondéNet, magazine publisher Condé Nast's Internet division.

**COOPER, BRIAN.** Director of new media, Ernst & Young.

**CORRIN, TANYA.** Former host, *TanyaTV*, on Pseudo.com, girlfriend of Josh Harris and cover model on *Nerve*'s first print issue.

**DACHIS, JEFF.** Cofounder and CEO, Razorfish.

**DANIELS, DEAN.** President and COO, TheGlobe.

**DAY, WILLIAM.** Former vice-president, Prodigy.

**EARLE, CAREY.** Cofounder, Word.com.

**EGAN, MICHAEL.** Alamo founder and TheGlobe investor.

**ELIN, GREG.** Motorcyclist, Silicon Alley to Silicon Valley, a TotalNY publicity stunt, 1995.

**ESSL, MIKE.** Former art director, SiteSpecific. Chief creative officer, Chopping Block.

**EVANS, NANCY.** Cofounder, iVillage.

**FIELD, GENEVIEVE.** Copublisher and editorial director, *Nerve.*

**FILO, DAVID.** Cofounder and chief yahoo, Yahoo.

**GRAZER, BRIAN.** Film producer.

**GREENE, BOB.** Partner, Flatiron Partners.

**GREENSTEIN, HOWARD.** Helped start WWWAC in 1994; later, "technical evangelist," Microsoft.

**GRISCOM, AMANDA.** Sister of Rufus and former intern, *FEED.*

**HALPIN, MIKKI.** Former editor in chief, *Stim,* a Prodigy-backed online magazine.

**HANSEL, SAUL.** Technology reporter, *New York Times.*

**HARMON, AMY.** Technology reporter, *New York Times.*

**HICKEY, CATHERINE.** Former Prodigy executive.

**HOROWITZ, DAVID.** MTV Networks founder and Alley angel investor.

**HUNTLEY, VANCE.** CTO, TheGlobe.com.

**JONES, SKIP.** Ran first White House website.

**JUDGECAL.** Former producer, Pseudo, and longtime Harris pal.

**LESSIN, BOB.** Chairman and co-CEO, Wit Capital. Former vice-chairman of Salomon Smith Barney.

**LIVACCARI, TOM.** Founding publisher, Word.com.

**MANNES, GEORGE.** Reporter, TheStreet.com, *New York Daily News.*

**MAYA, PAUL.** Cofounder and CEO, togglethis.

**MERRIMAN, DWIGHT.** Cofounder and CTO, DoubleClick.

**MCGINNIS, RYAN.** Artist represented by Razorfish Subnetwork (RSUB).

**MORITZ, MIKE.** Partner, Sequoia Capital.

**MURNIGHAN, JACK.** Former editor in chief, Nerve.com, and "Jack's Naughty Bits" columnist.

**NEGROPONTE, NICHOLAS.** Cofounder (1985) and director, MIT Media Laboratory.

**ODES, REBECCA.** Cofounder, gURL.com; graduate, ITP; and girlfriend, Craig Kanarick.

**PELSON, DAN.** President, CEO, and founder, Bolt Media.

**RAYMOND, STEVE.** Founder, *Stim*.

**READMOND, RONALD.** Co-CEO, Wit Capital. Formerly vice-chairman, Charles Schwab.

**RESSI, ADEO.** Founded interactive agency methodfive and sold it to Xceed for $75 million, mostly in stock, just months before the 2000 market crash wiped out 99 percent of Xceed's value.

**ROSSETTO, LOUIS.** Founded *Wired* in 1992, the first magazine to make cyber culture glamorous and a huge commercial success.

**SCARPA, MARC.** Cofounder and CEO, JumpCut.

**SHANNON, GABY (GABRIELLE).** Wife of Kyle and cofounder, UrbanDesires.

**STEIN, BOB.** Founder, Voyager.

**STEIN, LARA.** Former manager of creative development, Microsoft Multimedia Productions. *SAR* called the talent scout "the most recognizable Microsoft employee in New York" on its 1997 list.

**STITES, JANET.** Cofounder and publisher, *AlleyCat News*.

**SUH, CHAN.** Cofounder and CEO, Agency.com.

**SULZBERGER, ARTHUR.** Publisher, *New York Times*, and chairman, The New York Times Company.

**TEGE, JACQUES.** Cofounder, Pseudo.

**TOMLINSON, RAY.** Invented email and the use of the "@" symbol in email addresses while working on an ARPANET contract at Bolt Beranek and Newman (BBN), a Boston technology firm, in the early 1970s.

**TRIBE, MARK.** Founder, Rhizome.org, and another Brown alum.

**VAN METER, JONATHAN.** First editor, Word.com.

**WATSON, TOM.** Cofounder, @NY.

**WEINRICH, ANDREW.** President and CEO, sixdegrees.

**YARDENEY, HAGAI.** CEO, bla-bla.

# EVANGELISTS AND ENTREPRENEURS

*The rapid proliferation of the Internet and the World Wide Web in the late 1990s would not have been possible without the PC revolution of the '80s. But the story of the Internet is much more a tale of the triumph of network computing—the science of getting computers to talk to one another.*

*Back in the day when the world's most powerful computers took up entire rooms, and were controlled by punch cards and powered by vacuum tubes, scientists began looking for ways to access the country's half dozen supercomputers from afar.*

*In the late 1950s, in response to the Soviet Union's early lead in the space race and the paranoia of the Cold War, President Eisenhower set up the Advanced Research Projects Agency (ARPA, later changed to DARPA) to fund research and development of special projects for the Defense Department. One of its first tasks was the creation of a national computer network called the ARPANET to link these computers. The ARPANET would be a distributed network in which any computer could reach another computer though numerous paths on the network. Among the goals of the new network was to create a new way for surviving command centers to maintain communication with each other in the case of nuclear war, even if direct links between them were broken.*

*A series of technological breakthroughs in the following decades paved the way for ARPANET to emerge as the basis for a worldwide electronic communication network. Soon after, in 1969, Stanford and UCLA exchanged the first line of text over the ARPANET. TCP/IP (Transmission Control Protocol/Internet Protocol) emerged in the '70s network protocol as a kind of universal computer language; its worldwide adoption allowed local area networks (LANs, as office networks are known), to communicate with other networks over the Internet. By linking LANs to one another, the Internet became a network of networks.*

*In the 1980s, the Defense Department opened up the use of the Internet to scientists sharing information over the system. Through the late 1980s, librarians, academicians, and a small but growing collective of hobbyists were using the Internet. In 1991 the National Science Foundation helped make the Internet available to commercial enterprises for the first time.*

*That same year, a British scientist named Tim Berners-Lee, working at CERN—the European Organization for Nuclear Research in Geneva, Switzerland—released to the public a new system called the World Wide Web, which allowed users to view Internet sites as represented graphically in Hyper-Text Markup Language (HTML). The first Web users used FTP (File Transfer Protocol) to access CERN's computers and see the new visual Internet.*

*By the early 1990s, New York computer buffs had developed a handful of dial-up bulletin board services (BBSs) like Echo and SonicNet; meanwhile, a vibrant community of developers was working to turn the CD-ROM into the first successful interactive multimedia technology. The most visible of the CD-ROM developers was Voyager, a company that got attention for both its innovative work and its self-destructive management team.*

*But the advent of the Web displaced the promise of CD-stored interactive media, and soon the fledgling CD-ROM industry had all but disappeared. In its wake it left a budding group of computer graphic artists and digital entrepreneurs who helped seed the early Internet industry.*

*Another large talent pool emerged from the ranks of early Internet ventures like Prodigy, an online services company headquartered in White Plains, New York, that was victimized by America Online.*

*With the invention of Mosaic and Netscape, which enabled Internet users to browse graphical Web pages on the Internet, artists and technologists in New York began to experiment with visual effects and building some of the first websites in the world. Teaching themselves HTML, they soon mastered the science of coding for the Web.*

*And out of dinner parties thrown by Mark Stahlman and his Cyber Salon, the New York New Media Association (NYNMA) was formed, which helped build the concept of community through monthly events known as CyberSuds.*

**JACK HIDARY** | CEO, EARTHWEB | The people who created DARPA realized that in order to create a research community focused initially on nuclear research—because they were doing a lot of nuclear research at the time—they needed to connect their computers together. It was that simple. The first test bed was about four universities and two national laboratories. Stanford was in there, Columbia was in there, Los Alamos. So you had five, six different sites, all connected together. And it worked, and it also gave them a measure of redundancy, because if one node between Los Alamos and Stanford went down, you could still connect to Los Alamos another way.

This redundant system made things fundamentally different from the circuit-based switching we have in, say, our telephone system. The analogy I would use is the highway system. Imagine if I wanted to create a highway from here to L.A. just for me to drive on. That's basically what our telephone system is today, because you have to create one single circuit dedicated to your conversation between here and L.A. And that's expensive, just as a single highway between here and L.A. just for us would be. Instead, our highway system is very much like what the Internet is. We have on-ramps and off-ramps, but we all share.

ARPANET quickly grew and added university after university, because it was so useful and required very little maintenance. Each node maintained itself, which was the beauty of the system.

And then came TCP/IP—another fundamental difference from the telephone system. On the telephone system you use an open line, and the data that goes past has no address on it. If it gets routed somewhere else, nobody knows what to do with it. On the Internet each little piece of data is independent, and it can find its own way. It may take a while—you get garbled emails—but it will find its way to the ultimate destination. It also has multiple ways to get there. It's also easy to add on to. So the guy who first created the network doesn't have to decide beforehand who's going to be on the network, as you would in other kinds of networks.

**RED BURNS** | FOUNDER, INTERACTIVE TELECOMMUNICATIONS PROGRAM (ITP), NYU | I think it's one of the most wonderful ironies of all time that, because of the Cold War and because of ARPANET, we got ourselves a decentralized system that cannot be centralized.

I can remember in the early days hearing men at conferences saying, "Who runs the Internet? I want to talk to the man who runs the Internet."

**JACK HIDARY** When we used the Net as scientists, we ran it as a peer-based group. There was no head of the Internet, no chief honcho. It was peer-to-peer, groups of people designing protocols, and we all participated in that. The Net was difficult to use. You needed to know Unix, basically. If you were not comfortable with Unix, it was not a good idea to use the Net. We used the Net in science to transfer files to colleagues, to transfer papers, to transfer images. I come from that area. My specialization was brain imaging. Each scan was 5MB, so instead of sitting on a disk I would transfer on the Internet—and at that time it took longer than it does now.

We would transfer on the intranet at the National Institutes of Health. NIH has fourteen thousand scientists, just to give you a sense of the scope. No paper. We just emailed. The secretaries had email; everyone had email. So it was really one of the first communities to adopt an electronic way of life, if you will. I had international colleagues I corresponded with via email. Everyone had an email address. If you were a scientist and didn't have an email address, it was weird.

In 1991 came Tim Berners-Lee with the Web. He came out of CERN, which was a scientific laboratory. His initial provocation for creating the Web was for scientific papers—specifically physics papers—which have a lot of graphs. He wanted to show those graphs.

Also in 1991, the University of Minnesota created Gopher, which was basically a text-based menu system. Initially it was neck and neck; there were more Gopher sites than websites. People said, "Oh, cool, Gopher is faster and cleaner." There were thousands of Gopher sites. But then people started looking at the Web, and they looked at Tim Berners-Lee's very, very crude browser, which basically showed that you didn't have to deal with text, you didn't have to deal with images—just point and click. And he adopted it from something called SGML [Standard Generalized Markup Language], the markup language that textbook publishers used to mark up textbooks for publication. A lot of people think HTML is a programming language. It's actually a kind of metalanguage, a tag language. When a book goes to press, SGML tells the printer what exactly the font should be, what should be bold, etc.—just like HTML does today. SGML itself is somewhat complex; Berners-Lee simplified it for HTML. That was one of his geniuses.

**DOUGLAS RUSHKOFF** | WRITER | In the late 1980s the Internet was a play space, a public space. Commerce was actually illegal on the Internet. You had to sign an agreement that said you were using this for research purposes. And as a result, it was a very pure, Utopian, idealistic space. The people who were online had to observe certain rules to maintain this unadulterated, commercial-free zone. It was a place where we felt, for the time that we were online, free of that sort of survival instinct, free of that anxiety that comes about when you see commercials, when you're trying to sell, trying to make a profit. And as a result those of us who were online then imagined the spread of the Internet being literally the spread of a new cooperative global culture.

We had an inkling of what a global society might be like, of a very cooperative global organism, if you will, where ideas spread very freely and people cooperated in developing strategies for living together in peace. And people who were having these experiences online were having such a good time that they really wanted to spread this around.

**JACK HIDARY** Tim Berners-Lee is a great story, because the man created the Web, created everything we know of today, and yet decided to take no upside from that except for being head of the WWW Consortium [W3C], which helps set the standards for the Web.

Berners-Lee's vision of the Web was actually a lot different from the one we have today. In his vision, you would go—first of all, no addresses, no URLs—link to link to link. You would start with a link somewhere, and then just link to where you had to go. He never imagined the use of addresses by users themselves. He always thought that would be on the inside.

Second, he thought that every site would be two-way. You go to a site, you look at the content, and then you change the content. And the next time someone comes, it looks different. He thought this would be all interactive. It didn't turn out that way.

But in any case the Web was the breaking point. It was the mass point that took the Internet from a useful scientific tool to something every mom and pop can understand how to use. Remember, at that time, Al Gore and others were talking about the national information infrastructure—building a new highway system the way his father, Senator Gore, had built the actual highway system. And what ended up happening was, Berners-Lee just masked it—he put a front on it. He didn't change TCP/IP, the protocol that underlies the Internet. He said, Don't worry about all the TCP/IP; don't worry about the Unix commands. Just use the Web. Easy. Just point and click. That was a stroke of genius that ranks among, I don't know, Edison and the highest of high inventions.

One feels almost silly sometimes, not having thought of it. But we as scientists were used to the complexity of the Internet, and we assumed it had to be complex. I remember sitting at NIH the first time I saw the Web—my jaw literally dropped. I just sat there for ten minutes staring at it, and I knew that my life would never be the same, that *our* lives would never be the same. It became super apparent to me, as it did to many others, that this was a fundamental shift, not only in the way we were going to transfer information—because that's really the tip of the iceberg—but in our social and economic context as well. To anyone involved in it at the time, I think that was clear.

**JOHN BORTHWICK** | FOUNDER, TOTALNY | I first saw the Web at MIT's Artificial Intelligence lab in 1993. A friend of mine was studying up at the AI lab; I went up to see him and stayed with him for a long weekend in Boston. He was talking about the Web, and he said, "Come by and I'll show it to you." And he loaded up an early version of Mosaic on a Unix box and showed me the Web.

Most of my friends and peers who have grown up in Silicon Alley or Valley with this medium had transformative events when they first saw it. I just—it shook my world. It's like, this was everything I thought this network-based economy could become. It was already there, in very rudimentary forms. I sat there at the MIT AI lab and I went to a site a guy put out in France called the Web Louvre, which was one of the very early graphical sites, with a couple of images from the Louvre Museum in Paris. It was an unofficial site and eventually the Louvre asked him to take it down. But just to be able to sit down and realize you were actually seeing a page that was being served from a host in France, which an individual had put up, was an incredible experience.

**CRAIG KANARICK** | CO-FOUNDER, RAZORFISH | I've been using computers for a long time. I was in a chat room in 1979. I grew up in Minnesota, and there was this giant thing called the Minnesota Educational Computing Consortium—MECC. They're famous for having made this game called Oregon Trail. They had one giant mainframe—like the size of a couple buses. We had those acoustic coupler modems, where you dialed the phone and put the receiver in a little cradle. And it was on old teletype machines, with these giant paper rolls that would vibrate as you typed on them. Like the old military typewriters with the paper tape on the side that punched out little chads. People would dial in and use it as a programming resource, and there were chat rooms and there were multiplayer *Star Trek*–based games where you would explore the universe and try to shoot the other person. Our school was really into computers, so we got a chance to use all of this stuff. It was fascinating.

Then, in sixth grade, when the Apple II and the RadioShack TRS 80 came out, a whole bunch of my friends and I learned to play with those things. By the time I went away to college in '85, I had an Internet email address. People used to think I was crazy. I had friends who took a semester abroad, and I thought, instead of trying to call each other, let's email each other. If they weren't in the computer science department or the engineering department, they didn't know what I was talking about.

**JOSH HARRIS** | FOUNDER, JUPITER COMMUNICATIONS, PSEUDO. COM | I wrote the book on the Internet in 1990. The last report I wrote at Jupiter in 1990 was my best piece of writing on the subject: *Revenue Streams of Mass Market Video-Text*—as it was called back then—*A Case Study of the French Minitel System*. Basically, I had a guy in France get me all the numbers—at that time they were charging ten bucks per hour, and still might be—for Minitel. Minitel, back then, was actually a fairly vibrant medium: it was fairly modern, it had all the basic features that we have now, it was just kind of crude. But everything worked. So what I did was, I followed the ten dollars an hour they charged: where did every penny go? And after I did that, I pretty much understood how the Internet would unfold.

**RUFUS GRISCOM** | COFOUNDER, NERVE.COM | Steven Johnson [FEED], Nicholas Butterworth [SonicNet], Mark Tribe [Rhizome.org], and I all studied semiotics in college. We all had an interest in media, the way it works, and an awareness that power insinuates itself through technologies and through somewhat abstract forces. It made sense that this emerging new technology and its implications would show up on our radar. We'd been thinking along those lines for many years. We also had somewhat antiestablishment cultural leanings, and we didn't have good jobs, and we had been sitting there strumming our chins for three or four years out of college, and we were kind of ready to do something. It was good timing.

**CLAY SHIRKY** | FORMER CTO, SITESPECIFIC | At the time I was heavily into poststructuralist literary theory, as many people were. And this thing seemed like a poststructuralist wet dream. It was so incredible what was going on. When you name something, it comes into existence. Just the simple fact of life on the hard drive was to me really astonishing, because as a Mac user I'd only ever played in the shallow end of the pool. And there was a day where I was trying to FTP something from some remote site, and I wasn't getting it. Literally pounding the table in frustration. And I was like, Okay, I have to think through this. How does this work? Oh, I get it. Directories are kind of like folders on the Mac. And then it hit me—that, in fact, everything the Macintosh had told me about computing was a lie, and the truth was the text. That all of the pictorial stuff was a representation of directories and files and not vice versa. It's like that moment where you discover that *Romeo and Juliet* is not a rip-off of *West Side Story*.

**STEFANIE SYMAN** | COFOUNDER, FEED | If you look at the founders of that first wave, we're all pretty close in age. Not a coincidence: we all graduated into a recession, so none of us had really attached ourselves to any career path in an unchangeable way. None of us was looking at rising up through the corporate structure.

The other creative outlets were not so attractive in the middle of a recession. You just had a feeling there was going to be nothing but endless shitty work and not very much pay rather than something more glamorous—*I'm going to be a writer, and by my second book I'm going to do this, or I'm going to be a filmmaker.* It was a different landscape.

We were all kind of comfortable with technology, having had computers at a much younger age than previous generations. And we all recognized that something was happening, while the people with the power at the time, who were above us, had no clue. I remember my first boss being scared of a Mac. And I remember going to organizations run by people who are now our age—young thirties—and thinking, this whole office needs to be networked. Why isn't it networked? It seemed so obvious, but people weren't even focused on those issues, let alone on how to do it the right way.

**STACY HORN** | FOUNDER, ECHO | I attribute the fact that Echo survived to Gore and Clinton, because just when I was starting to think, This is not going to make it, they started talking about the "information superhighway," and it changed everything. After that, it was like there was this national guilt about the Internet; everybody felt, *Oh, I should try this, or I'm going to miss the boat.* Suddenly, when I would go to a cocktail party and start talking about the Internet, instead of people saying, "What's that?" they would say, "Oh, yeah, I've heard of that. That's something I can try. You're the Internet? Okay, I'll try you."

Then the media jumped on it. And because I was local and I was a woman and I wasn't a geek and I could talk about it in human terms, I got all this press for a while that really helped my business grow.

**OMAR WASOW** | FOUNDER, NYONLINE | When I started NYOnline in 1993, I originally had this idea for a black online community. I was going to call it something like Diaspora.com. People were like, "Diaspora? Die-ass-poor? What is that? You want to make an online community for you and the other three black nerds in America? We don't get it." It was pretty clear for me early on that we needed to shift from something that was focused on African Americans to focusing on a diverse online community. That was partly straight economics and partly the experience I had growing up in New York City. Trying to create an online community that was as heterogeneous as the New York I grew up in. When I was building NYOnline in '94, people weren't talking about an urban Internet, an urban online. People weren't even talking very much about the Internet.

**TIM NYE** | FOUNDER, SONICNET | SonicNet, in the early days, was like everything you would imagine a start-up to be. People in my loft till all hours. Punk hair. I remember trying to kick these skinhead punk rockers out of my apartment at one A.M., and they wouldn't stop the chat session, and I wanted to go to sleep. It was so fun; in any other industry there was such a history that you felt like you were doing something that plenty of people had done better. Not here. You were writing rules, and trying things, and there was such uncappable potential.

Nicholas [Butterworth] was in a tight group of friends of mine. We had passed each other through a lot of mutual friends and had been friendly since we were fourteen or fifteen, but never had connected. I was asking around for someone who could really help me conceptualize, and I brought Nicholas in as a consultant, and we got along like a house on fire. I knew SonicNet needed someone who knew what was cool musically, but also someone who had very good instincts for interactivity.

**NICHOLAS BUTTERWORTH** | FORMER PRESIDENT, SONICNET | I think when I got to SonicNet the BBS had four modems, meaning four people could connect simultaneously and chat with each other. The service we offered was basically your own email account, with a Sonic-Net.com address. We offered access to Internet services like FTP and Gopher, chat with other people on the BBS, message boards where you could leave messages for other users and they could pick them up later when they were on. And then we had a project to build directories or databases of album art photos, bios, and sound clips from unsigned bands and independent bands. Tim's original idea for the service had been as a business-to-business thing, where bookers to small clubs could find out about bands that were available. And there was also a ticketing function with a unique number, and then someone at the door would have a list.

Then we did a digital download project with the band Future Sounds of London, which came out at the same time CompuServe did a download of an Aerosmith song. I think it took about an hour and a half to download a three-minute song. A couple hundred people did it. Very early on they had hit on some key applications for music and the Internet. There was another company called IUMA, which was pure Internet, but those were basically the two Internet music services that existed as far as we knew.

So you had these national services—Delphi, Genie, CompuServe, AOL, and Prodigy—and they seemed like the big players. But they weren't so big; they all had less than a million subscribers, I'd guess. We thought we could compete and create a niche as dedicated audio.

We had given out a bunch of comp accounts to people in the music business in New York, artists included. And we started doing a lot of chat programming, where we would have artists come online and chat, which we still do; we've hosted thousands of chats with artists. For lots of people it was their first encounter with online and what it meant.

I think we added four more modems. At the high point I think we had three or five thousand subscribers in New York paying ten dollars a month, but I don't know if everyone really paid. I think a lot of them were comp accounts, and I don't think we were very good at collections from the people who didn't pay. It was mostly people from the indie rock scene, and they weren't exactly diligent bill payers.

In '93 the protocol was written, and by the end of '94 there was this new company called Netscape, formed by this guy Marc Andreessen, who had written Mosaic. They put out this product called Mozilla—it had a little animation of a monster—and that was the first Web browser I ever used.

The Web seemed like an interesting service, but just one of many Internet services. You had Telnet, FTP, email, and the Web. So we started trying to figure out how to add it to our BBS, and also how we could use the Web as a gateway to let people dial into our BBS remotely. Our plan was to establish little SonicNets in other cities. We thought local scenes were really important, so we were going to have Boston, New York, Chicago, San Francisco, and link them all somehow, so you'd be a member of your local BBS and they'd all be linked. We were wrong about that. We were wrong about the Web and what it was good for. We thought since we were charging ten dollars a month to access our BBS, and the service they were getting was email access, Internet access, and content as a bundle, we figured that the content, which was the most important part, must be worth at least four bucks unbundled. So our plan was, if people came to the website, they could subscribe to the BBS for only four bucks a month, but if they want to dial up with email, that would cost them ten dollars.

**JACK HIDARY**    I had a rush of adrenaline when I first saw the Web, and I just got a bug that I had to be involved. I told my boss, and he understood and supported me. I said, I don't know what I'm going to start exactly, but I need to go do it. I hooked up with my brother and my friend Nova, and the three of us got going. We realized that it was too early to sell a product; it was too early to start a mature business model. The one thing we knew we could make money off was consulting. So we used that as an interim step toward our final goal. The first year and a half we did consulting; our first big client was the Metropolitan Museum of Art.

**MARISA BOWE | FORMER EDITOR IN CHIEF, WORD.COM |** For me, the Web felt like having sex with ten rubbers on. Whereas in a dial-up environment, you really feel that psychic energy of other people.

I'd been doing Echo in a pre-Windows environment. I had this computer with this endless blue background and no Windows, so it was just the text against this deep blue space background.

**JOHN BORTHWICK**   When we started up TotalNY, which was one of the first local sites in the city, we were pitching the fact that it's available twenty-four hours. "Twenty-four hours, seven days a week, at a fraction of the cost." It's just interesting to see what we thought was important back then. "Twenty to thirty million Internet users world-wide." You notice we were saying Internet users. There were only four to five million World Wide Web users worldwide.

**MARK STAHLMAN** | COFOUNDER, NEW YORK NEW MEDIA ASSO-CIATION | I retired from Wall Street in '92. After I brought America Online public, I didn't have to work much anymore. I guess I began to have some withdrawal pains. There was an article published in the *Utne Reader* about people pulling together salons. And so I decided to start a salon. We called it the Cyber Salon, and I'd say probably one to two hundred people passed through. It ran for close to three years, once a month. It was a monthly dinner party for people who were interested in computer-related, network-related activities in New York, at a time when people were actually leaving New York because they couldn't figure out how to build a business here. I had a number of friends who packed up their families and their companies and left, went to California.

**JASON MCCABE CALACANIS** | FOUNDER, SILICON ALLEY REPORTER | When I graduated school, it was a recession. You were lucky to get a job. When you got out of school you might be on the job circuit for six months, twelve months. I don't know if anybody can con-ceive of that today—like being lucky enough to get an interview. You were excited if you got an interview, and if you got a job for twenty grand a year you took it and you were happy. You were careful not to upset anybody, because if you lost your job you might be on the street for six months or a year.

Everybody I talked to said, We're not hiring. We're not hiring. "Can you get me an interview?" I didn't work for two, three, four, five years, and everybody would say, "Well, what is your experience?" And I'd say, "Well, I don't have any experience; I've never had a job. I've only worked at a café or bar." I don't know if people graduating today even remember that time. But it was this whole Catch-22: everybody was saying, "I can't get experience because nobody will hire me."

You got one week of vacation and you were lucky, and you didn't take it, probably, and you worked hours and hours and hours of overtime. And you didn't expect anything out of it, and you never had in your head that you'd ever be a millionaire. It was an impossibility. The way you became a millionaire before the Internet was, you won the lottery.

**RUFUS GRISCOM**   I was a slacker after I graduated from Brown in 1991. I took a year off and was a ski instructor for a year, a ski bum. We had limited ambitions; we basically felt there were no attractive jobs. Maybe the best-case scenario was that you get to be a journalist and slave away on a terrible salary, not getting to write what you want, or you worked at a publishing house under similar circumstances, or you sold your soul and basically exchanged your twenties and thirties for a chunk of cash on Wall Street—which I considered, begrudgingly.

All the options seemed quite grim. The best possible option seemed to be to make coffee at a local diner and write, or to play in a rock band—which is what Nicholas [Butterworth] did after he graduated from Brown, and what I did to a degree, and what a lot of other people I know did. When I graduated, the brightest people I knew didn't even leave campus; they just got a stupid job in Providence, Rhode Island, and hung out. These were the brightest people around.

It was a very, very different environment. In many ways, though, it was a wonderful one, because it was kind of like Prague before the Velvet Revolution. There was something attractive about being an intellectual dissident, especially in the aftermath of the '80s, and I think that's been taken away from the younger people today—in the sense that there used to be a paradigm where you could either make money or do something interesting, and that made not making money appear very attractive. In fact, not making money was a symbol of your integrity.

**DOUGLAS RUSHKOFF** Nobody from everyday workaday reality believed that what we were doing meant anything at all. I used to get laughed out of cocktail parties in New York City in 1994 for suggesting to people that they would someday be using email. My first book, *Cyberia,* was canceled by Bantam Books in 1993 because they felt the Internet would be over by the time the book came out. They thought it was like CB radio, and the book industry was just too slow to be able to get it out in time before the fad was over. I was shocked and horrified. It was largely New York not wanting to believe that something cool could come from the West Coast. You know, Kurt Cobain went and shot himself. Grunge would die, the Internet would die, this was your psychedelic crowd, this wouldn't happen here.

**JASON CHERVOKAS | COFOUNDER, @NY |** New York City had suffered through the impact of the '87 stock market crash, followed by the early 1990s recession. It had lost something like 450,000 jobs and hadn't had an organic, homegrown, grassroots, growth industry since television left for the West Coast in the early 1950s. You have to picture the city when we got there. There were really not a lot of job opportunities for young, talented people. So you had all these sort of underemployed creative types, for whom New York had always been a magnet to begin with, sitting there without job opportunities. You drop first-generation HTML in their lap, which was designed to be a publishing platform for nongeeks in academia. They latch onto this as a publishing platform for these naïve, fantastic dreams. They thought they were going to be media moguls.

**JASON MCCABE CALACANIS** It was a renaissance, and it was all these people who were out of work and couldn't get jobs being able to dream about something and be empowered. You know, imagine not being able to find a job, and then all of a sudden being able to start a company on the Web and not having to ask anybody's permission. It was incredible. It was like the shackles were lifted. *We can be our own bosses. We can do something that's never been done.* It was truly, truly inspiring. And that's what the parties were really about. They were about people talking and brainstorming and sharing their experiences: Have you tried this? Have you seen this website?

**MARK STAHLMAN**  As the salon progressed, as a lot of topics were discussed, a couple of us—Brian Horey in particular and I, but a lot of other people also—figured it was probably time that we sort of extended this whole process. The first thought we had was, Let's open this up to the public. So in effect the first activity of this dinner party was to make a bigger party. In fact, I think the original mission statement of the New York New Media Association was to "galvanize a community in New York." I remember that word *galvanize* being debated over, and a variety of things being proposed.

**CELLA IRVINE | FORMER HEAD SIDEWALK.COM NEW YORK |** In '94 Brian and Mark had this idea for a meeting, and a bunch of us volunteered to be on the board of NYNMA, and we had the first CyberSuds at El Teddy. Fifteen people showed up, and they kicked us out because we weren't worth serving—we weren't worth giving their private room to. We didn't make them enough money. And now we get about three thousand people.

**DAVID LIU | COFOUNDER, THE KNOT |** My partner Carley [Roney] and I had a CD-ROM development company. There was an arrogance to the CD-ROM companies, because they followed more of a publishing model. There was a sense that this was for the intelligentsia—*Let's create these very esoteric things.* Voyager was a major culprit. And what I was most disturbed with, and the reason we jumped over to the Internet even though we were a profitable business, was this: There were a lot of people who were going to pay us a lot of money to create CD-ROMs. But we realized that as media people one of the big tests we had for whether or not a medium was going to work was whether or not you become a consumer for your own work. We had a staff of twelve people in the office, and not a single person went out and said, I have to buy this new CD-ROM; let me pop in a CD-ROM and experience it. And when that happens you realize that the media itself is flawed— something is wrong. Yet some dumpy little CompuServe account that was sitting in the corner would always get accessed every day, because people would always be checking their email. Human behavior was going to dictate where the business was going to follow.

**THERESA DUNCAN** | CEO, VALENTINE MEDIA | I started out making CD-ROMs, and I kept making them far after I should have stopped. I studied linguistics, and one of the principal reasons I was interested in multimedia was the difference it would make to narrative. So I actually did some writing about how it hearkens back to oral story-telling—it's an epistemological change in the way people learn things. You learn through your whole body in multimedia. Through your eyes and your ears and your sense of movement. And for me that was really exciting. The direction that I initially thought the Internet would go hasn't really fulfilled its promise, though.

**DAVID LIU** The Internet brought on a whole lot of people who were a lot scrappier than the CD-ROM people. The Internet was more a matter of pure information, because you didn't have the bells and whistles to leverage.

We really believed this was going to be a new pipeline into American homes, a direct pipeline that was empowered with interactivity and applications that could really catalyze a lot of other things, including transactions.

Moving from CD-ROMs to the Internet was actually a step back, when you think about it. One of the things we worked really hard on with our CD-ROM company was to create a highly visual and dynamic experience, one that was less hierarchical. Instead of having a menu and making you choose from a whole list of things, we were trying to figure out how to create something that was more of an interactive slide show, more like interactive film. But the best you could do at the time depended on the bandwidth of the CD-ROM. The Internet back then depended on 2,400-baud modems, and the big debate was, Is the 9,600-baud modem really going to catch on? Because if so we can actually add in some graphics. That was a difficult transition.

**MARC SINGER** | FOUNDER, TOGGLETHIS | In '94 the Internet wasn't cool and multimedia wasn't cool. I was working at Times Mirror, a newspaper company, making CD-ROMs, but they weren't one of the leading CD-ROM makers at all—they weren't even really interested in it. But it ended up being brilliant for me, because Paul Maya, who I started toggle with, was also looking at the Internet, and we were both thinking, The Web's going to be great—we've been doing this CD-ROM stuff, which is really cool, but the CD-ROM industry is a hard industry to get distribution for, because shelf space is really hard. I don't know who the big people were, whether it was Electronic Arts or Microsoft. But if you didn't know somebody, if you didn't have a good distribution, you couldn't get your product out there. We saw that the Internet could solve the distribution problem, and we saw we should just start our own company to do it. It was so early that no one got that this was going to be interesting.

In many ways the Web is a much higher hurdle to jump on the entertainment or marketing side than CD-ROMs were. Because in the CD you had a massive thing that you could put all this content on. Trying to put it on the Web was altogether different. On togglethis, we were trying to re-create the CD-ROM experience, or at least parts of it, without having this giant CD-ROM capacity to work with. The Web is a lot harder in some respects, because back then it was mostly non-art-directed Web pages, and no one was thinking about creating a strong emotional experience for anything.

In 1995 I was a writer on staff at another company doing CD-ROMs. When our company got a big project, I said, "Wait a second, this could be cool." Everyone knew you could do really cool interactive multimedia stuff, whether it's for big corporate pharmaceutical companies or advertising agencies or whoever. But no one's going to see it, I realized, because you can't get anyone to buy the CD-ROMs.

The Web's great because you can actually get people to see it.

**GENE DEROSE** | CEO, JUPITER COMMUNICATIONS | When I first got to Jupiter, the decaying Zeitgeist was around Voyager. Almost from the moment I got here we were aware of it. Josh [Harris] always believed that online was where it was going to be at, and really was singular among people I knew at the time. CD-ROMs and interactive television were talked about just as much as online at that point in time. But Josh always pooh-poohed Bob Stein [of Voyager] as just being totally past it, or missing out on it.

*Why did Voyager fall apart?*

**GENE DEROSE**   There was never a business sense there. It was about quality art and quality content, and executing that in the wrong medium, it turned out, was a huge mistake. It probably was as simple as that. I don't know Bob Stein, but I've heard some things that made it sound like he was also just iconoclastic to a fault. Which so is Josh Harris, arguably, but it's a matter of whether you're surfing the Zeitgeist or it's low tide. I don't think the company had any identity outside of Stein, and CD-ROM was not going to be anything that captured people.

**CLAY SHIRKY**   We assumed that, within the world of the Web, national barriers meant nothing. News was sneaking in through Usenet (a public bulletin board system) that the Canadian courts were trying to shut down the Internet. We were just like, Look, it's wired across the borders, there's nothing you can do. And then these guys would come in waiting to write us checks, but their legal departments were breathing down their necks about it. It was interesting to see these kind of early battle lines, which still haven't played out. But the astonishing thing to me, and the thing that just imprinted—the way it would imprint into an infant—was that the power of the thing came from the low barriers to entry, which meant that amateurs could become professionals. Jerry Yang and David Filo over at Stanford had this little thing called Yahoo, and people went to their site. *Two guys.*

Meanwhile, everyone in the commercial world was trying to use the Web as a kind of low-grade example of how great everything was going to be when we finally had professionally produced CD-ROM content. Bob Stein at Voyager was banging on about that. I saw the guy from *Blender* [a CD-ROM magazine] get up and give a talk at one of Jaime Levy's birthday parties, and making fun of the Web because it was so slow, and saying that *Blender* was really going to be the future because it had all this rich multimedia content on CD-ROMs. And I remember looking at him, thinking, You're a moron. It recognizably isn't right. And so many people had so much money committed to interactive TV this and CD-ROM multimedia future that, that we actually had tons of breathing room, because no one would take us seriously.

**JASON MCCABE CALACANIS**   When I say the Internet [industry] is only five years old, I am talking about the commercial Internet from '94 in the fall, when Mosaic started spreading images. Which for me was the starting point: the starting point of the Internet industry was when Mosaic first put an image up. In everybody's mind, every advertiser in the world, every marketer: *This medium can sell things.* This medium can be used to market. In text they didn't believe it could— text, despite how the literati would sort of take it, is not the medium of choice for the overwhelming majority of Americans. People are into visual stimuli and audio stimuli, and that's where most people want to consume stuff. People don't like to read, is the bottom line; and if they do they want to read in very small chunks and they want the information organized in some very tight fashion. And that's when the Internet became, I would say—the commercial venture.

**MARC SINGER**   People were skeptical of the Web. The first couple of years it was like "This is a fad, this isn't going to work." Wall Street people said, "This has nothing to do with us." Ad agencies said, "This is not advertising." Hollywood studios said, "This is a fad, this isn't going to work."

In the beginning of Silicon Alley, the pitch always began with, "Here's the Internet." Because most of the people had no idea what it was. The average person you were trying to do a deal with, the average investor, had no idea what it was. And then when you showed it to them, by the time they understood it, they were seeing a static Web page. So there was a huge disconnect. And there still is. Whether you're Warner Brothers or individual writers and artists, the Web gives you the opportunity to interact directly with your audience, one on one, which people are still not fully taking advantage of. So most of the corporate sites look pretty much the same.

**DOUGLAS RUSHKOFF**   There was a desperate struggle in New York to figure out: How do we make money off this thing? There are a lot of people having a lot of fun talking to each other off keyboards. How do we turn that into a moneymaking opportunity? So the first thing that happened, from '94 to '97, was: We're going to create online magazines. We're going to create FEED and Spiv and Stim, these places where people can chat about things, and read things, and slowly try to get them to care more about our expert content than their own stupid airy-fairy conversations. And most of them failed, because no one wanted to sit there and read that stuff online.

**RUFUS GRISCOM**   There's really a very limited number of actual content brands that emerged in the Internet space. There was a window of time when everyone was thinking, "Oh my gosh, there's this sort of fissure in how the entertainment industry distributes content and brands and information and ideas. This is an opportunity for us to inject new ideas into the media." There was a lot of countercultural enthusiasm about that opportunity in '95 and '96—and then basically people concluded it wasn't a good business model. And since then you don't hear about kids in college saying "Wow, I wanna start a Web magazine." People don't say that.

**ANNA WHEATLEY** | COFOUNDER, ALLEY CAT NEWS | Once people started thinking of the Internet commercially, the rapidity with which companies started in New York will never cease to amaze me. This is not a hospitable town for entrepreneurs. And if you want to get into publishing, your parents have to bankroll you. Because coming out of college you're making seventeen thousand dollars, and even if you have five roommates it's still tight. In '94 or '95 this was a version of getting an MBA.

These were a lot of people who were either right out of school, or their families were willing to support them. Maybe spouses were initially less willing, which is why you have such a pronounced younger set of entrepreneurs at the start. There was a sense of *let's give it a try*. In '94 to'95, it all solidified; *New York* magazine came out in November or December with its "Cyber 60."

The term "Silicon Alley" first appeared in the *New York Times* and the *Wall Street Journal*—so therefore it existed. People started to identify with Silicon Alley; the New York New Media Association was a focal point, and started growing and doing outreach into the community. People argued whether or not they represented the community. They really were a visible element.

**MARK STAHLMAN**   I came up with the name "Silicon Alley." I think it came to me in a dream—these things usually do—and I went to the board of the New York New Media Association, and I sort of ran it past them to see if they had any problems with it, if they saw any conflicts or whatever. I was cofounder of that organization, so I knew what was going on with that, and they liked the name. So I gave it to them, and that's how those things happen. I'd previously coined a few names like this. I had a ten-year career as a Wall Street analyst and banker, so I'd made up a few things along the way and they seemed to work. "Network computing" was actually something that was pretty successful in its day. And actually I'm kind of surprised at how well "new media" has done, which I also made up.

**SCOTT KURNIT | FOUNDER AND CEO, ABOUT.COM |** I hate the term Silicon Alley. Silicon Rock, which is the term I prefer, is strong, rather than some crummy little offshoot of Silicon Valley. To me the name "Silicon Alley" is one of the worst things that has happened to Silicon Alley since it was formed. Because it comes off as a little stepchild to the Valley. So maybe we should call it Internet Rock or Silicon Rock.

Jason [Calacanis] wouldn't be happy with that, since he owns the URL.

**JASON MCCABE CALACANIS** Everybody thought San Francisco was going to be the Internet capital of the world. That was the primary thinking. Every Internet company would be based in San Francisco, and that would be the "new media" capital of the world. That was the term everybody used in the media—which I never liked, because how long could it be new? There are a lot of problems with the term "new media," so I never really use it. I always used "the Internet" or some other term.

I never liked the term "content," either, because I always said in the early days: content is a way of venture capitalists commodifying art. You can't commodify art. If you came up to me and said, "I like your content," I'd be sort of insulted. Just like if you went up to Bob Dylan and said, "Gee, I like your content," or "Michelangelo, I like your content." It's like, Okay, you like my widgets. It's not what writing or music or photography or video is about. It's art. And media.

And I thought to myself, Gee, New York is the capital of so many things: finance, art, advertising . . . and that was when I realized—advertising is going to be big on this thing. And this is before they were really bannering, but I knew it was going to work, advertising, and I knew that Wall Street was going to be involved in this because I'd witnessed the PC wars and experienced the sort of spoils of it, getting paid a very high salary for a very young person at the time [building LANs as a college student].

So I realized that finance and advertising were the two biggest things. All of finance was going through Wall Street. I heard a statistic early in my career that seventy percent of advertising dollars that get spent in the United States go through Madison Avenue. And to me that cinched it—New York was going to win in terms of the application of the Internet. And I made my gamble. I said, Okay, I'm going to cover the Internet in New York.

**THERESA DUNCAN**   There was a lot of excitement, there were a lot of young people, but I don't think anybody had any idea of the IPO madness or big money that was to come.

**JOHN BORTHWICK**   I started up my company [TotalNY], started bringing together the team. In October or '94, *Wired* ran a story—which wasn't a cover story, it was in the inside of the issue—called, "The Next Revolution Has Begun: The World Wide Web" ["The (Second Phase of the) Revolution Has Begun," *Wired*, October 1994]. I remember sitting there at lunchtime, on my own, reading with sweaty palms. I was thinking, People and media companies are going to read this—the secret was out. Now everybody was going to be on top of this thing. So I felt like I had like months—not quite weeks, but months, if that—to build up the company and stake out some territory.

**JASON CHERVOKAS**   It was culturally fascinating. I remember going to a TotalNY party in the summer of '95 and thinking, this must be what San Francisco was like in 1966, before the world discovered that there was an actual subculture and something happening and something to productize. There's something beyond just business going on here.

**NICHOLAS BUTTERWORTH**   When TotalNY launched, that's when I met Ed Bennett, at that party. Ed had just come on to run Prodigy. He came from MTV Networks, he'd run VH1—he was a media guy, not a tech guy. He talked about how Prodigy was going to be Web-centric instead of proprietary (giving access to Prodigy content only), and those of us who had realized the Web was the greatest thing thought that was very important. He talked a lot about content and how that would be differentiating. I had worked with MTV at Rock the Vote—we were a content play, we were a music site, we were working with artists. We thought there was an audience for hip-hop online.

**JASON CHERVOKAS**   Prodigy had a much bigger footprint and a much bigger impact in its time, in terms of employing people, training people, investing in Pseudo and SonicNet. And it was also very short-lived. A lot of people passed through Prodigy's payroll department. And the same thing with this very short-lived, never-launched MCI/Delphi thing. Everyone worked at Delphi. It was one of the big three with Prodigy and Pathfinder as things that seeded the business side, and they cast a long shadow.

**SCOTT KURNIT**   I was there during a very interesting time in Prodigy's history. It had already spent one billion dollars, which was a huge amount of money back then. I'd been there for just under a year and a half when we turned it profitable, and a lot of people had gone. So there were a lot of people who came before and after who I don't really know. I'm not sure Prodigy turned out a lot of entrepreneurs, because it was a pretty safe haven funded by IBM and Sears. It's probably the antientrepreneurial environment.

When we started About.com, we did recruit a dozen of Prodigy's best and brightest into the organization, and stopped only after getting a letter saying that we shouldn't be doing that.

**JASON CHERVOKAS**   Pathfinder, Prodigy, and MCI/Delphi seeded Silicon Alley with talent and taught people to do what they do. And to a degree Voyager, but Voyager burned very brightly, very briefly, and threw off a lot of creative people. Those four businesses really trained people and threw their creativity into this mix.

**NICHOLAS BUTTERWORTH**    None of us at SonicNet were company people looking for a company. It's just, this project seemed like— we thought we were going to totally change music forever, and we thought we were going to empower a whole generation by doing it. It was just incredibly exciting. When we started thinking about our website, we figured we could invent all kinds of rules for what that would be, and new paradigms for content and design.

We had a weird way of starting. In May, when we switched our focus to the Web, instead of launching one site and sticking to it, we thought we would have to get good at production, so instead we planned a series of sites through the summer of '95. And our discipline was such that we would launch a new website every two weeks. So we had a team of four people working on these websites. One was about Coney Island—sideshows by the seashore. One was a fictional account of these two aliens. We had this slogan, "Loser friendly," so these were, like, the aliens from the planet Loser that had come to New York. We went out and shot digital video of them doing all this crazy stuff in costumes, and made stills and animated gifs from it. That was called the "Losers' Guide to New York." So it was this series of almost episodic magazinelike content features. We got good at HTML and production process, and then we launched the website and switched the BBS off forever in the fall of '95.

**CLAY SHIRKY**    I remember when every day there was someone willing to get up and say, "This Web thing will never work. It's not going to threaten our business, we're going to have nothing to do with this." And you're laughing.

Because as long as you're channeling what you know to be true about the nature of the online business—it was like having bullet-proof armor for a while there, it was so fun. You just walk into meetings, and people would be yelling at you: "This is going to threaten our business." Okay, see you! It was like being able to walk through walls.

People at AT&T would drive in from New Jersey to meet with a dozen twenty-five-year-olds in the living room of an apartment in Koreatown every Tuesday for a day—engineers, marketing people— and we'd sit around this metal table and we'd try to think how we, the twelve of us, were going to help AT&T, the world's oldest telecommunications company.

People you couldn't have gotten a meeting with in three months if you'd gone through the channels would come to see us. We later realized they were coming in to see us rather than having us go out to Basking Ridge because Basking Ridge was so dull, and they were kind of getting a kick out of the adventure, too.

**FERNANDO ESPUELAS** | CEO AND CHAIRMAN, STARMEDIA | I was a user of the Internet from 1994, when I moved back to the States from Latin America. I had been working in Argentina for Ogilvy, and although Argentina made incredible strides in the 1990s it was still relatively backward in terms of technology, even in 1994. When I came back I was posted in Miami, and there was not only the Internet, but also obviously email. I started to use email for business, which I thought was an incredible tool.

About a month before I came back, we were filming a commercial in Buenos Aires, and it was raining. When you're about to film a commercial, there's a moment every day—call it nine o'clock—when if you decide to cancel the shoot you have to pay for the shoot regardless of whether you film anything. They call them weather days. So there's an enormous amount of pressure. To make a long story short, I tried to get in touch with my client, who was ten blocks away. I could not get a phone line to connect to my client ten blocks away. I literally ran across the city to get the client to tell me to cancel the shoot. From that to email—that's the contrast I saw and felt. I understood fundamentally that what the Internet represented at some level was not just another technological wave or something that brought convenience, but really a complete shift in how these societies could be organized and how people in these societies could view themselves.

**SETH GOLDSTEIN** | COFOUNDER, SITESPECIFIC | People from different walks of life, with nontraditional backgrounds, who didn't have jobs or had part-time jobs, all working together on something that was totally new, that nobody knew how to do or where it was going— that's unique, that was special. Whatever technology comes next, whether it's wireless or broadband, nobody really believes it will be as disruptive, as radical, as the Internet's emergence was as a commercial medium in '95. Communication media platforms like the Internet only come around once a generation.

**NICHOLAS BUTTERWORTH**  It was an amazing time, when you felt like you could invent something and create it for the first time ever. And you could bring together incredibly creative people and figure out something new. Designers, technical people, musicians, writers, and all imagine what this stuff is capable of and then do it—and every one of them could be the first.

In '95—especially that year—you could see the medium exploding and developing every time you went online. The experience of discovering things all the time—which totally expanded your ideas of what was possible—was just exhilarating. And I feel like that was shared by all the people that were involved.

# IF YOU BUILD I.T., WILL THEY COME?

*E*ntrepreneurs *who were turned on to the Web in '94 and '95 quickly taught themselves HTML, or HyperText Markup Language, the standard source code for building Web pages. HTML allows designers to use text commands to make visual pages that can be accessed by any Web browser. Though HTML is actually relatively simple, and not even a true programming language, its power to turn text into rich media made it a language treasured by its speakers. Early devotees quickly realized they had the power to do something for which wealthy companies would pay dearly. Suddenly, twenty-five-year-olds were gathering their friends together, opening start-ups in their living rooms, and preparing million-dollar pitches for AT&T, Duracell, and IBM.*

*If this was the Gold Rush, early interactive agencies were the ones selling the shovels. Between 1995 and 2000, almost every major corporation in America caught on to the significance of the Web, but most had to outsource the development to an interactive agency, and contracts to build new sites ranged from a few hundred thousand to several million dollars. Almost overnight, an unexpected and unprecedented new market had been created, and the race was on to seize the cascade of new business.*

*With all the money to be made, design shops cropped up quickly, hiring new workers in droves, and evolving into full-service consulting companies, evangelizing the tenets of the new economy to their newfound corporate devotees. The mantra at Razorfish, perhaps the brashest firm in the Alley, made the direction of the future clear: "Everything that can be digital will be."*

*Razorfish—started by Craig Kanarick and Jeff Dachis, a pair of childhood buddies—and Agency.com, started by Chan Suh and Kyle Shannon, an ad salesman and an actor who met online, grew up in parallel in the Alley.*

*1994 and 1995 were an especially inspired years for Kyle Shannon, who launched three bold ideas in three short months: Urban Desires, an online magazine; the World Wide Web Artists Consortium (WWWAC) a support network for Web designers; and, Agency.com, a website design firm.*

**KYLE SHANNON** | COFOUNDER, AGENCY.COM | We started Urban-Desires, the World Wide Web Artists Consortium, and Agency.com in a three-month period, and all three of them took off. All three were like "Boom!" And Agency really took off—it was a rocket sled. I'd put in three months of sweat equity investment just in concepting these things, and then pulled a few triggers, launched a few things, and I was sucked into a vortex, moving forward so fast I was truly holding on.

**JASON CHERVOKAS** In the early days there were the Web service shops, which all sort of started as would-be media companies. Agency really grew out of UrbanDesires; Razorfish really grew out of what was then The Blue Dot (an online art collection), or they grew up together. The idea of all these guys—Kyle and Chan, Jeff and Craig—was that they were going to be media moguls. They were going to build these media properties, and they were going to design websites for corporate services to pay the freight in the short term. It also just so happened that because Madison Avenue was here, website design was the first business sector that developed as an actual cash-flow-positive business. That was an accidental thing; I don't think it involved much planning at all.

**CRAIG KANARICK** My first impressions of the Web were *God, HTML is easy,* and *This is never going to work.* You have limited control over how it looks, and it's going to look a little different on every screen, which is crazy. I thought, this thing a great tool for cross-linking physics papers, but I don't know about anything more than that. That was on the one hand. But on the other hand I thought it might be huge. It was schizophrenic. There was optimism, but there was also a healthy dose of cynicism. I had seen these things before; it was still clunky and ugly, and I just couldn't imagine something that ugly catching on. But Windows is ugly as hell and it caught on, so I was a little naïve. I didn't think about business forces and what clever marketing can do.

I was excited that it was a little bit easier than Gopher or Usenet or FTP or command-line interfaces and older things. I thought it was fun, so I thought other people would find it interesting, too. I just didn't think the way it was set up was going to work. When things like CGI [Common Gateway Interface] came out—which gave us the ability to do real computing on the server, like forms or credit card processing— then I knew it was a much larger opportunity than just static text documents. That was when we really started to get some confidence. That was where the motivation for doing all this came from. We thought, Man, another really fucking ugly computer application that's going to try to spread around the world. Why can't people make these things better-looking? Well, that's what I went to graduate school [MIT Media Lab] for, *I* know how to make this stuff better-looking, I'm going to get in the business and make really beautiful digital experiences. I thought, Good design will win out in the end, because people will realize that it's important and will hire us.

**KYLE SHANNON** Chan [Suh, cofounder, Agency.com] worked at Time Inc. for nine years; I had a less illustrious career. We were talking about figuring out ways to make money doing this content stuff. He'd been ten years in magazine publishing, and said the publishing business is a really shitty business. The good ones lose money for five years straight. A lot of magazines were losing money—sometimes millions per year—and I thought, that's just insane. In those days we didn't have access to IPOs, we didn't have access to VCs [venture capitalists], and we didn't have a dime to rub together between us. We knew we had to have a business where we could get the money first and then start the business. So we thought, sell the services first.

**CRAIG KANARICK**  Razorfish started because of selfish reasons and opportunity reasons. I was doing some consulting work in this space, designing interfaces. I was freelancing for companies and doing CD-ROMs and anything digital. It wasn't about the Internet per se. I was consulting for a small company in Connecticut; they had been in interface design for eight years. But I hated schlepping up to Westport every couple of days. And I thought, If these guys have four people after eight years, they're only going to have four people after another eight years. And this opportunity is huge.

**KYLE SHANNON**  Chan had been doing Vibe Online and trying to sell some online advertising to the *Vibe* print advertisers. He said, "For an extra ten grand I'll give you an advertisement deal on the website," and they're like, "That's great, but we don't have a website to put there," and he said, "Well, we could build it for you." So he built a couple of websites that became content for Vibe Online. There was one called the Timex Map, which had the time zones of the world, and you clicked on them and up came a Timex watch with the hands in the correct position for that time zone. For 1995, this was pretty advanced stuff. He said, "I have a feeling that these big companies might be willing to give us money to do what we do." And I, being an actor, said, "Okay, I'm ready to quit my day job."

**CRAIG KANARICK**  In the beginning of '94 I was still consulting, and I was still helping people at *Blender* magazine, which was trying to be a CD-ROM monthly music magazine, or a little bit of everything. I was coming in there trying to see if they had a job for me, because I was really sick of freelancing.

I thought I'd start a company. Instead of just being an individual person, I decided to pick a name and print up some business cards and be the only employee of the company.

**KYLE SHANNON**  We talked to the guy from the *Sports Illustrated* swimsuit video, and they had a TV show coming up on the fourteenth of February and it was already the end of January. They said, "Can you make us a website?"

We said, "Sure, just give us the content."

They said, "Well it's not quite done yet."

**CRAIG KANARICK**   I grew up knowing Jeff [Dachis, cofounder, Razorfish]; we met in nursery school. When we both went away to college we didn't keep in touch. We were acquaintances; I wouldn't say we were great friends. But we saw each other probably three or four times a week, because we went to Hebrew school together and youth group and camp, and had mutual friends. So even though we saw each other all the time, there were probably always lots of other people around and it wasn't like the two of us hanging out.

I was still coming in to *Blender* once or twice a week and helping them unpack computers and install them. Jeff was coming in there also, to talk to them about their marketing needs and the possibility of hiring him to head up their marketing activity. They realized that we were both from Minneapolis and said something to Jeff—did he know this guy Craig Kanarick? And he said, "Sure." Or so the story goes.

**KYLE SHANNON**   So Saturday night, at four in the morning, a week before the show's supposed to air, they hand us a videotape and say, "Okay, boys, go make your site."

We were like, "What do you want it to be?"

They said, "We don't know—you're the Web guys."

They had sold off all the picture rights, and we only had a hundred and twenty seconds of video we could use. It was just bare bones, but we got it done and launched it, and it had huge downloads, and we crashed the Time Inc. server. Luckily we had backup servers, because we knew the Time Inc. server would crash. And it got enough press that it really got Agency started, and from there it just snowballed. We got some work with the Time Inc. consumer marketing division, who gave us some swanky offices by the loading dock. They always smelled like gas fumes, but they were offices. Then we had offices in the Time & Life Building, and they leased us machines at a reasonable rate so we could have machines when people sat down.

**CRAIG KANARICK** I come home one day and Jeff's on the answering machine. Here's a guy I haven't talked to in eleven years calling me up and saying, "Hey, I hear you're trying to start a business in this new digital world, and I'm trying to start a new business. Maybe we should get together and chat."

I went to his apartment at seventh and Avenue C, a house that Geraldo Rivera used to live in. The East Village wasn't as gentrified in '95 as it is now. I went over to his place and showed him Mosaic. We talked about the Web, and he talked about his direct marketing and guerilla marketing experience and having just taken his brother's financial services company public in Minnesota. We talked for a couple of weeks and that was it. It was great; it was a combination of wanting digital experiences to be better, and Wow, there's a huge business opportunity, too.

**KYLE SHANNON** Chan resigned, but they let him continue doing work at Time Inc. because very few people knew how to do what he was doing. Chan architected the beginnings of what would become Pathfinder [Time Inc.'s first magazine website], but he didn't like the direction that was going. Vibe Online was going fine, but then I stumbled into the picture and I started pushing pretty hard for him to quit his day job. All of those things added up to him saying, "Fuck it, let's go."

**CRAIG KANARICK** We sat around and talked about all the possible stuff we could do, and we didn't know whether it was going to be the consulting business that we're in now, the content business, or the software business. We just knew digital technology was going to be huge. Our first website actually had three divisions listed: Razorfish Solutions, Razorfish Studios and Technologies, and Razorfish Consulting, which is the one that took off.

**KYLE SHANNON** When you start your own business, the first people you approach are people you know. So we could walk through the halls at Time Inc. and knock on some doors, which is what we did. Our first clients were *Sports Illustrated,* Time Inc. Consumer Marketing, and HBO Home Video. From those we got Hitachi, Columbia House, Met Life, Amex, GTE . . . all in a three-month period. Then we got an

RFP—request for proposal—from GE, and at two o'clock in the morning we decided not to pitch the work, because we realized if we had gotten that work we wouldn't have been able to do the other work.

**CRAIG KANARICK**  Everyone asks how we named the company Razorfish. I used to say I needed a really great story—like, "We were in the Caribbean, and we were shaving, and there were fish." We played the naming game. I was giving lectures and still doing some odd jobs here and there, and we both had other things that we were doing, but we knew we had to name the company.

Then someone at Time Warner called me up to hire Jeff and me to do a website for them. When we came in and talked about our capabilities they said, "Great, what's the company called?" We hadn't decided, but they said, "Well, we want to hire you by next Friday. By next Friday we're going to cut you a check to give you the up-front half of your payment, so you have to tell us who to make the check out to and what your EIN number is, or we're going to have to pay you in cash, and I don't think you want that."

Every single thing we looked at became a possible name. We specifically wanted a name that didn't describe what we did—that was going to get people to say, "What does that mean?" We didn't want a name like Agency.com, or like CyberHyperSolutionsIncorporated, that would lock us in to something specific. We also wanted to make sure that we were differentiated from different companies. Were you Digital Solutions or Digital Solvency? There was just too much confusion in other things we saw.

The word *razorfish* came out of flipping through the dictionary. I remember sitting in Jeff's living room flipping through every book we could find, just looking for any words at all. We had reams of paper with words on them. We had a phone that wasn't plugged in that we would answer fifty times in a row, practicing answering the phone with that name and seeing how quickly we got sick of it. That crossed off tons of them. Once we were down to ten or twenty names, we just sort of said, "Razorfish? Okay, let's go with that."

**KYLE SHANNON**  We spent the next six months doing the work we landed in those first three months. And hiring. And it was—like everyone talks about in a start-up—seven days a week, until you literally can't do it anymore, and you just go home to get enough sleep to take a nap and come back and do it again. For seven or eight months it was just hell on earth.

**CLAY SHIRKY** I was making these little Web pages and learning the nuts and bolts of HTML, because I was coauthoring a beginner's guide [for Ziff-Davis]. I had a real professional's attitude toward the Web, and I was getting so mad that all the stuff out there was so bad and amateurish, that in a fit of pique I wrote "The Worst Web Pages of the World Wide Web." My goal was to set the bottom tier for Web design and make something that was didactically bad—leopard-skin background, blinking yellow text, everything I saw people doing. So I put this thing up, and a couple days later I get mail from a guy named Kyle Shannon, who says, "That is the worst piece of shit I have ever seen. Would you like a job?"

I was going out with a woman on Echo at the time who knew Kyle Shannon, and she said, "Oh, those guys are hard-core. You should definitely talk to them."

**CRAIG KANARICK** I gave a lecture at Time Warner to their internal communications teams. They had a little conference and I remember this lecture incredibly well because a couple of interesting events happened—not the least of which was that Chan Suh gave a talk at that same event showing the Vibe website that he had set up without permission, and was getting a lot of grief from people in the room. I remember talking about what the Internet was and what it could do, and Chan talking about looking for advertising and making people nervous.

**CLAY SHIRKY** I got to Agency.com, and my training in theater had been as a production manager and stage manager, so my title at Agency was *integrator*—what does that mean? No one knows. But it ended up meaning "production manager." Just tried to get everybody working together, because there were *no* content management tools. The only content management tool was force of will. As long as you remember to do things the same way twice, you'll be okay.

I probably spent the most time working on the initial Columbia

House site, which was really, really instructive. We were totally on fire for the Web, and these Columbia House people were incredibly stupid and slow. Ultimately, the site didn't launch because what they wanted was great, big marble buttons with drop shadows, and we were trying to advance the art of interface design.

I spent a lot of time with Columbia House. And that led me to understand all sorts of things about the cultural clash between the old economy and the new. They were running on a big old database, and they had no way to update everything for the Web all at once. So I could see, though I didn't understand at the time, how much it would come to mean to me. It was the difference between one company founded to take advantage of the Web on Web-like principles, and another company trying to back its way slowly into the Web.

**KYLE SHANNON** Chan and I were in early, and we knew we were pretty respected. It's sort of like a club: once you start working with Fortune 500 companies, others will at least look at you because they know someone else is trusting you. We knew we were in that club. You can get kicked out of the club by doing shitty work, but we were at least in it, and that was good. We knew that if we didn't get kicked out of the club, and we played our cards right, because we were one of the early players we had a chance to be one of the significant companies as this Internet thing evolved.

**CLAY SHIRKY** We were producing UrbanDesires out of Agency and using it as a kind of design lab. Kyle and Gaby [Kyle's wife, Gabrielle Shannon] were running UrbanDesires, and the people at Agency became the designers and producers of UrbanDesires. I did a lot of interfaces for the art section. Everything had these innovative interfaces, which I later realized completely obscured the presentation of the art. We were so totally excited about Web pages as their own thing that we couldn't understand what other people wanted.

**LISA NAPOLI | REPORTER, MSNBC |** There was a venture capital media meeting downtown. Chan Suh and Kyle Shannon were at the event, talking about business models, and you could tell they had never talked about a business model to a financial crowd before. It was kind of cute and disarming. I mean, they are funny guys, but they clearly were not in the same universe as these suited bankers—and at that time it didn't matter. Everybody was so impressed with the fact that these two guys happened to know about the Web that there seemed to be forgiveness.

**CLAY SHIRKY** Over time I became the CTO of Agency. That period was when the term Silicon Alley was invented. In the beginning—1995 and 1996—there was Agency, SiteSpecific, Razorfish, and I think FunnyGarbage. There were a couple of days when RealAudio launched, and you could hear poor AM quality voices over the Web, and it rocked our world. And when Razorfish launched The Blue Dot, we were sitting around looking at it and were just . . . up until that point, in our minds, we had been absolutely cock of the walk. We knew of no one else who was doing design as well as Agency. The Blue Dot came up, and we wanted to hate it, but we looked at it and said, "Wow, this is really good."

They [Razorfish] had really taken the kind of server push gif animation style—before animated gifs existed—and they'd created this kind of weird visual landscape. I don't really remember any of the pieces, except for the moving blue dot. We just had our asses kicked, and we knew it. Suddenly there was enough competition that everyone was upping the stakes all the time. There were a bunch of us and we all did Web stuff. You know, whatever you want: You want the Web? We do that. Come on down, give us a bag of money.

Over time people started to "get big, get niche, or get out," as they say. Agency went into the Fortune 1000 world. Razorfish specialized, relatively, in new launches, and SiteSpecific really became a marketing company, because we had an in-house media department long before other people did. We all started in this pool, and we all separated and stretched out the ecology at the same time. [Scott] Heiferman launched a media company way before—I remember Kyle pointing him out to me and saying, "That guy just quit his job with Sony to go start an online media company." We never even thought about the Web as media, and Scott was already selling it.

*Scott Heiferman was twenty-two when he started his own agency, i-Traffic. A few short years later he would sell the company to Agency.com for fifteen million dollars in cash and stock.*

**SCOTT HEIFERMAN** | FOUNDER, I-TRAFFIC | I had quit my job at Sony, where I had met Chan Suh. This is when Agency.com was just getting started in the Time & Life Building. They had this office with no windows in this little closet space. There were eight or so people. I told Chan, "I'm starting a company that's going to do this," and he invited me over just to hang out one evening in that office. I remember being there, and it was the most amazingly cool thing ever, and all it was was just a bunch of ugly cubes in the middle of this ugly building. But the fact that there were these—it sounds so dumb now, but the fact that there was this group of people banded together to do something in the Internet world, and that they seemed really busy, it was so. . . . I didn't have any thought in my head that i-Traffic was going to be an agency. I'd never seen an ad agency before. I'd never physically *been* in an ad agency. You might think someone who starts a company has great vision, can see things he never saw before—screw that. I saw Agency. com being a half-dozen people; that was our model.

I went out for pasta and pizza that evening with the company. I had just left Sony Electronics in New Jersey, where there was no concept of going out to eat and talking about your work at eight P.M. And that, for a twenty-two-year-old kid for whom this was the most interesting stuff on the planet—that was totally a turn-on.

**OMAR WASOW**   Chan was working at Vibe, and I went out to this consumer electronics show in Chicago. Chan is out there trying to hustle up some advertising deals, and he says, "Why don't you come out? You can crash in my hotel room, and I'll get you some food. You have to pay for your own ticket, but you can work the booth, and you can hustle a little and I can hustle a little." It was totally seat of the pants, scrappy. I go and work this booth a little, and it's like, you've got these people with *castles* for booths, and millions of dollars' worth of the latest Sega something-or-other. On the bus back from the show I sit next to a guy who's working at Sony—Sony has a little electronic organizer—

and that's Scott Heiferman. So we have a friendly chat, and I think maybe he's going to comp me one of these Sony things and put me in the beta test group. We hit it off, and a couple years later I bump into him when he and another friend, Seth Goldstein, are sharing the same loft space. Their office was an apartment next to a "massage parlor."

**SCOTT HEIFERMAN**   I used to say that an entrepreneur shouldn't spend so much time thinking about capital, because I remember seeing the Jeff Dachises and Chan Suhs raising so much money from Omnicom and making friends with bankers, and I thought the whole idea of playing the Wall Street game was unnecessary. What I was experiencing, successfully, was this idea that you could grow, grow, grow, grow just by having good customers, providing a good service, getting results, having happy customers and a happy team serving those customers, and growing and growing.

I'm not saying a company has to be bootstrapped or ignore the need for capital. But the idea of a leader investing a lot of his time in the game of press or the game of Wall Street or the game of politicking for capital, I didn't understand. Then, when it became clear that these Agencys and Razorfishes were zooming in growth, and acquiring, and going public, I still believe I was doing the right thing for me and for i-Traffic, because I wasn't sophisticated enough even to understand that whole world. But it was clear at that moment that there was a level of sophistication in guys like Chan Suh and Jeff Dachis, businesswise, that I wasn't on par with.

**JEREMY HAFT | COFOUNDER, SITESPECIFIC |** I think Scott was the original germ that got us all interested. He's a good guy, a funny guy, but also very visionary in many ways. It was unbelievable that he was so young. I still tease him and say he's twenty-four. He was always such a lad.

**KYLE SHANNON**   We used to joke that the Internet would be as big as CB radio. CB radio was huge, and then they wrote a hit song about it and it disappeared—you never heard about it again. Was the Internet going to be one of those? Were we going to be the kings of CB radio color handsets? One of the surprises from that early vision was that the Internet didn't go away. Even though every news story today is about the demise of the dot-com world, it ain't going away.

**CLAY SHIRKY**   I ultimately didn't stay at Agency, partly because my role there as production manager—and as a geek-to-English translator for the clients—had become more valuable than what I really thought I wanted, which was to be a Perl programmer. So with some sadness I went into Kyle's office one day and said "This isn't working out," though I continued to work as a freelancer.

**CRAIG KANARICK**   We knew all the people in the other offices on our floor. There was a lawyer down the hall and a designer. Early on, when we had a client come in, we had one of the neighbors stick his head in and say, "Those documents you were waiting for—they're ready now." I think we may have once said that Ryan McGinnis, who was next door, worked for us. He's a painter, but also a graphic designer, and he ended up doing a couple of design jobs for us. The reason we did that was people didn't believe that we could actually do as much as we did. We'd told that client it was just the six of us doing all this work, but he couldn't fathom the fact that we were staying there twelve hours a day, every day. He thought we must have other people working for us. So we sort of said, "Yeah, we have some consultants down the hall who help us out, too." It wasn't this big deception to make us appear bigger than we were in order to win the job. I think he was afraid that if in fact it was only the six or eight or twelve of us doing all that work, that we were going to explode. So to calm his fears we said we had this support system.

We had no air-conditioning. We had this crazy thing called the MovinCool, which is a portable air conditioner. It was this giant machine, with tubes that went out the window. It was basically a dehumidifier; it condensed all the moisture in the air into these big tubs, and we used to have to carry the ten-gallon jug of water to the bathroom and dump it out and then put it back. The thing would break all the time and then it would be two hundred degrees.

Once we had a meeting, with Lara Stein and a bunch of people from

Microsoft when they were first starting to come to New York looking for content. We had some meeting in the conference room, and all of a sudden sewage starts pouring out of the ceiling. Dirty, brown water—I think it was rust. Some pipe had burst, and here we are in the middle of some meeting and water starts pouring down in the middle of the conference room table. I'm sure it happens to everyone.

**DOUGLAS RUSHKOFF**   A big turning point was when Microsoft came to town in '97. They set up shop in Worldwide Plaza, and Lara Stein came to fund all these projects. They wanted MSN to replace AOL, and they were going to throw all this money into content. And really what I think they were doing was just trying to get people to use Microsoft tools rather than the Mac, and to use Microsoft Explorer rather than Netscape.

So they hired some artsy people to convince all the artsy people in New York that if you do a Microsoft project, you'll get all this money, and you'll get all this artistic freedom. I don't think they ever really had an intention of creating that world, of using that content. I think they just wanted to get people to use their tools. So they gave seed money to a lot of different people, and created this series of milestones. You would do this first part and get more money, do the second part and get more money, and I think it was just a way to get us to buy.

Any of the money they gave you, you ended up spending on PCs and Microsoft products anyway, to get up to speed in their world. Then the projects all were canceled. But by then people were developing with PCs. That was an interesting moment; it really turned New York development culture from Mac culture to PC culture—which is a different culture, a different thing. Once you turn in your Mac for a PC, it's like you've taken that first step into the Borg. Then assimilation is inevitable.

*In 1996 the Omnicom Group decided to "play the agency sector" by buying significant portions of Agency, Razorfish, and other shops, placing a shrewd bet in the growth of the industry. Its initial investment of $11.7 million in Agency, for example, was worth $1.1 billion after Agency's December 1999 IPO.*

**ANNA WHEATLEY**  Omnicom was one of the smartest investors, because it was the only one that said, I don't care which one is moving—I'll take a little Agency, I'll take a little Razorfish. The whole Razorfish/Agency model—these people have been successful, but I don't think it can be sustained. You just have to be too hip to be cool enough to work there. That personally bothers me, because I think that that is stupid business. Again, I might be naïve, because they're going to clients and selling themselves on the fact that they are that cool.

**KYLE SHANNON**  I knew if the industry survived we'd have a good chance of being a leader, but I had no idea what being a leader meant or looked like. If you'd asked me five years ago if we were going to have more than sixteen hundred employees, an office in Korea, Sydney, four in Europe, and ten in the States, I would've said, "No, stop smoking crack." But that's where we are.

*Agency.com went public on December 10, 1999, closing its first day of trading at $62 per share, transforming Agency into a $2.2 billion corporation, and giving Kyle Shannon and Chan Suh a net worth of nearly $300 million each.*

**CRAIG KANARICK**  There was an interesting set of different goals. For me it was about winning an award for being the best place to work, or winning a design award. And for Jeff it was about being a successful company and making a lot of money. I think a lot of companies, when they start up, don't have that balance. They're one or the other. We had a really great balance, with each of us having interest in what the other did. Even though I had degrees in computer science and design, I had a natural aptitude for running businesses and understanding business issues. And even though Jeff was running the business, and doing that side of it, he had undergraduate degrees in performing arts and took grad school classes and was a dancer and a designer and ran a marketing company and understood communications. There were overlapping circles, as opposed to having lots of people in one camp and lots in the other.

*Razorfish went public on April 30, 1999, and the company quickly increased in value. By the end of 1999, Kanarick and Dachis's start-up was worth $4.7* *billion, and the two were worth more than $220 million each. Kanarick reveled in the media spotlight, changing his hair color every month and his suits to match, buying classic cars off the street on a whim, and commandeering an ice cream truck to rescue employees from a sweltering New York summer day. The founders flew around the world buying up companies, and Razorfish soon employed nearly two thousand people.*

**THERESA DUNCAN**  There is a sort of divide in Silicon Alley between the liberal arts majors and the business majors. But at Razorfish they were liberal arts majors, and they're the most vicious, you know, nineteenth-century capitalists. It's really funny. A lot of people remind me of those legendary nineteenth-century capitalists, the kind who were caricatured as a bag of money.

**JASON CHERVOKAS**  There was a sense then that [website] developers were in charge. It was a creative time for developers, and it spurred a lot of creativity, but it was a lousy time for product, because people did things because they could, rather than because anyone would want it or use it. In the days when the 14.4 modem was state-of-the-art, people were designing huge image-map navigation structures for websites that no one could possibly use, except for the handful of people—and I mean *handful* of people—who had any access to bandwidth whatsoever. There was a real disconnect. No one was thinking about the end user, who they were, what they wanted, and where this was going. They thought, this was cool, let's make it and see what happens. Which was exciting—it brought a level of creativity to what we're doing that is very much missing today. But they thought, let's throw this against the wall and see what sticks, and we'll all be worth a billion dollars tomorrow.

*SiteSpecific was an interactive agency started by Seth Goldstein, Jeremy Haft, and David Byman.*

**JEREMY HAFT**  In 1994 a college buddy of mine named Seth Goldstein was living on my couch. I was a journalist, freelancing for the *Observer* and *Quest* magazine, writing histories of New York families and shit like that. Seth had turned me on to the Internet, and I remember in the beginning how much it sucked and how unimpressed I was. But there was something hypnotic about it that pulled me in. I remember the first experience sitting in front of the computer with Seth and David Byman. Byman was this crazy genius nappy-haired guy; he was working at [Vibe Online] doing programming for their site. He created some recording-artist program where you could sort of create your own band. It was being demo'd to me, and I remember thinking, "Gee, what a dud." But once I started looking beyond the immediate limitations, it got very exciting.

It's funny that a lot of people from the arts went into the Internet. I think the early Silicon Alley companies were very much like theater companies. They were ensembles of people who learned to play to each other's strengths and weaknesses and worked all the time. There was a very artsy feeling about it in the early days.

The Internet has provided a personal odyssey and journey for a lot of the people that went into it. It became a canvas to refashion and repaint your life. I still think the Internet provides you with a way to form a new trajectory and your own personal path. Look at how people reinvent themselves every six months. New job, new discipline. Clay [Shirky] was in lighting with the Wooster Group, and he wound up teaching himself technology on the side.

SiteSpecific was an ad hoc event. The Duracell online business was suddenly open for pitches. Seth was working at CondéNet while we were pitching Duracell, the story goes, and he was fired on the day we won the account because they discovered he was working on a pitch for another company.

**SETH GOLDSTEIN**   It was a calculated risk; there were some people in the office when we were there working late on the Duracell pitch, and they weren't too happy about it.

**JEREMY HAFT**   Ogilvy had apparently walked in with purple boards. Clearly they didn't know how to communicate the Duracell brand if they were coming in with purple, because Duracell's brand is black and copper. We found a contact who knew the brand manager for the big five, which are AA, AAA, C, D, and 9-volt. We assembled a happy band of freelancers. There was Seth, Dave Byman for programming, and Mike Essl, who was an art director; I was writing copy and creative directing with Mike and a few other folks.

We basically built a site for them as the pitch. It was a pretty spot-on prototype, strongly branded—black and copper. This was all instinctual for us, because what did we know? Turns out they dug it; they hired us; they wrote us a check for two hundred and fifty thousand dollars. We incorporated and bought computers and we were in business. We launched our first website, Duracell.com, in November of '95, and then proceeded to really innovate in terms of the media. Duracell was really one of the first advertisers to be using banners on Yahoo, HotWired. None of the banner sizes had been standardized in '96. Our idea was to make it look as though the page were being penetrated by a Duracell battery. So there was a rip in the page as though it were paper and the battery nub was sticking out. If you clicked on the battery, you saw the reverse side of the page that you were just on. So if you were on Yahoo you saw Yahoo flipped, and a battery pack of Duracells and the line "Powered by Duracell." From there you could go back to Yahoo if you chose or you could go on to Duracell. The idea was that all the great sites on the Web were powered by Duracell. This was not a banner that took you to the website to drive traffic. This was a media-specific event.

That's how SiteSpecific was started. Then we proceeded to learn marketing on the client's dime.

**SETH GOLDSTEIN** We all came out of the theater, Clay and Jeremy and I. There was a sense of casting. You cast your technical person, you cast your creative person, then you show up for the client and everyone has to look their part. So much of it was about packaging and confidence building. It's one big confidence game, and it's still unclear where the real business is, in terms of profits. Until that's clear, it's really just about making people feel that you know more than they do.

**JEREMY HAFT** We moved into 1234 Broadway, which is on Thirty-first Street and Broadway. It was the apartment of Seth Goldstein and Scott Heiferman. After Seth moved from my couch he moved in with Scott, who was forming i-Traffic.

This building was hysterical. It was basically a hovel, except for a few apartments they were renovating so they could charge higher rent. Down the hall from us was an escort service, and on the other side there were bicycles all over, because there were Korean bike messengers that lived there. You can imagine bringing clients up into that building. Duracell was coming in, and it was just creepy. It was a strange building. We lived there until the SiteSpecific folks and the i-Traffic folks were practically hanging out of the rafters. There were two sleeping lofts; half was Site and the other half was i-Traffic. That's where Pascaline, Kevin Ryan's wife, actually worked at Site. The sleeping loft was four feet high, so everyone was always hunched over. We stayed there until we were putting desks in the refrigerator.

There were some great times in that space. We had won some AT&T business; we were building the HomeTown Network, which was their foray into content and community. The whole division was eventually lopped off, but the AT&T folks—about five or six guys—would always come on Wednesdays because we had a West Indian housekeeper who cooked lunch for everyone on Wednesdays. We would meet in the morning and then we'd all have lunch.

When we were pitching clients early on, and we wanted to bring them to the office, it was like central casting. Seth would call all his friends and say, "Come in and look like you're working." So we would all arrive fifteen minutes before the client would arrive, and would be sitting at our desks typing away. That theme that you find among these early start-ups was that there was a degree of theater to it. People were to a certain extent performing and acting and pretending; you put up the shadow puppets, and Look! We're a company! For some companies that's worked to their benefit, because the advertising business is theater anyway in terms of sales and presentation. It was a good skill to have.

There were shifts going twenty-four hours a day. Dave Byman would begin work at five in the afternoon and stay and work all night. He had programmers who would prefer that, too, so he had a nocturnal shift. Seth and Scott lived in the apartment. I remember one night when it was four A.M., and we were working on a big proposal for Duracell to up the retainer and trying to create a pitch for the following year's business. We just went over to Seth, because we were practically in his sleeping loft. He would be sleeping on the futon, and whenever we had a question we'd just wake him up. We just lived there. It was very much like an ensemble cast. These casts work together, grow together, live together, spend all their time together—that's what it was.

**CLAY SHIRKY**   Slowly, real money came into the industry. But unusually, everyone from stem to stern in the organization *knew* there was real money. In the advertising world, only the top ten people in any given company know how much money is involved. Everybody else is just working for their salary. But in the Web world, because it was a fairly tight community, everybody knew, everybody could feel it when the money started arriving. And one of the things we were always onto at SiteSpecific, for good or ill, was the sort of business-plan-of-the-month club; there was so much opportunity that we couldn't help but jump after this thing or that thing. And in retrospect trying to diversify ended up really damaging our overall valuation, because it turned out that building websites, even today, is a very profitable business.

**JEREMY HAFT**   In '95 and '96 none of us had any idea of the magnitude of the money that would come in. We never thought we were going to make a killing and retire; it was more this evangelical quest. There was something about the hypnotic pull of the Internet. It turned us all into apostles. For the first two or three years there was nothing else I could talk about. People used to make fun of me and Seth. We were very tight; everywhere we would go—parties, whatever—it was all we could talk about. We ate and breathed it all the time. Our vision of where the industry would be was much more idealistic. It wasn't a money thing; it was, "Hey, here's this pure channel." We might as well have been debating the Declaration of Independence or the Rights of Man. There was real fire. It was the passion of the arts.

**SETH GOLDSTEIN**   A lot of my favorite memories had to do with getting into meetings with people you'd never get into. We went to pitch John Deere. I don't even know where they were—Indiana? Michigan? But we all went to the John Deere corporate headquarters, and went through the John Deere tractor museum, and met with John Deere people and talked about the John Deere website. We went to Kmart, Sears, a bank, a hedge fund; we were all twentysomethings in really bad suits just geeking out, and really confident about what we were saying because we knew it better than anyone else.

**JEREMY HAFT**   By the end of '96 the country started to take notice. Suddenly there was more of an embrace by media, there was more coverage, and the pace of founding start-ups accelerated. Once we started to hear clues from the outside world that people were starting to pay attention, it felt more real. For a while we didn't feel real, we just felt like it was a crazy-ass job.

**SETH GOLDSTEIN**   I remember the moment Netscape introduced tables [a feature that helped site designers place graphic elements more precisely]. That was exciting. It was Chan [Suh] and David Byman and myself with a couple of other people. We all huddled around the computer and were like, "Whoa, this is pretty cool." Every incremental feature Netscape added to its browser meant a whole new forum for interactive communication, and that was what we were excited about. It wasn't about the great new world or about new media, it was really more about how the tools were getting better to allow us to do more interesting, more exact things with the medium. It was a craftsman mentality; we were building these things with our hands. There were really no off-the-shelf software tools to host a secure Web server; all that stuff had to get built from scratch.

**JEREMY HAFT**   The thing about Seth is, in some ways he's a maverick. He's very persistent and has this extremely impressive dose of self-confidence. In terms of business, he also had unbelievable self-discipline. This was the staying power that engaged us all early on and kept us focused and moving.

**SETH GOLDSTEIN**   It was really exciting, working with friends, trying to hustle and find people who knew people who could produce, program, design, or write copy for a website. It was very haphazard, there were no rules, and we were all making it up as we went along. That was the high. There was really a sense that we were all digging up fresh earth.

We spent time going down to the Bowery picking out steel tables with roll wheels. David Byman, who was a cofounder of SiteSpecific and the head programmer, had very particular work habits, and he said we needed to have Aeron chairs, no matter what. His cousin knew the distributor. I think we were the first Alley company to have Aeron chairs, which have now become commonplace. We had parties every month with a theme, and did Web-based invitations. It was kind of for new business, but also just celebrating who we were and the people we knew and people we were working with. We had parties called "Pasta & Perl Scripts," "High Tekka Maki," "Show Me Your Cookie," "Enchilogon." It wasn't really arrogant, just fun. There was a kind of flexibility and sense of irreverence. It wasn't about the money. It wasn't about popularity. It really felt like the playing field was level, and the only thing you needed was your creativity and street smarts.

**JEREMY HAFT**   Shirky had unbelievable personal habits. When we were in Little Korea, Clay would eat his kim chee using pencils as chopsticks. We'd be like, "Shirky, you're going to get poisoned." And he'd say, "It's not lead, it's graphite. It's fine, it's fine."

**CLAY SHIRKY**   A real turning point in our company was when Seth fired Sue Boster [Site's Director of Marketing and Goldstein's friend since high school]. Sue was one of the founding members; she was part of what gave the place its culture. And you can talk endlessly in those situations about who's right and who's wrong, but the bald fact of the matter was something wasn't working between the two of them, and Seth happened to be CEO instead of Sue, and he was the one who got to do the firing. It was handled very badly. It was done late at night, kind of abruptly, and Sue was told not to come in the next day. So she suddenly went from being around all the time to being persona non grata. It was a real moment of shock for the company, and suggested that we were a business, and there were a lot of people still resistant to that.

**SETH GOLDSTEIN**   Some time in '97 it changed. But from mid-1995 to the end of '96, it was magical. In '97, as the IPOs started to hit the market, it became more of a real business, and the investment bankers started to leave their jobs to go to dot-com start-ups and venture capital started to flow in: '97 to '99 was really about the money. Now it's about the morgue and the funeral.

**JEREMY HAFT**   First we had a minority investment from Harte-Hanks Communications. Harte-Hanks is one of the world's largest direct marketing companies. They do vast mailings; they have a huge database capability. Their CEO and president were both on our board, and proceeded to kick our ass and teach us about the model. I mean, our model of optimization and measurement was birthed out of the direct marketing model, obviously, and these guys were the ones who taught us. That's how we got the 3M account, and we sold through to a number of Harte-Hanks businesses.

There was a real culture shock. Harte-Hanks was a Texas company. They're, what, ninety years old? Originally in newspapers. They were Texas suit-wearing good ole boys. I remember—and this is funny—there I was in jeans and raggedy hair, talking to Larry Franklin, CEO of Harte-Hanks, and he says, "You know, you have a face made for radio. Son, son, you got a face made for radio there."

There was an article in the *Journal* that came out about the site and about Seth and about the deal and about the investment. And there was a lot of frustration. Richard Hochhouser of Harte-Hanks had been interviewed, and there was—there was frustration on their side. None of these Web businesses were real businesses. We were still coalescing. So all of the processes you would find—business processes that were so well-scripted in other businesses—weren't in place with us. We were sort of making it up along the way, trying to figure out how can we sustain growth, and build revenues, and maintain our cost structure. We were coming at it from this improvisational, intuitive side, and they were saying, "You guys are our investment, and you aren't delivering the revenues we had hoped." So there was a real clash.

About '97, we were looking for another round of financing. It was originally going to be a minority investment, and we were talking to a lot of people, especially Interpublic. We were talking to a lot of Interpublic guys, and the Interpublic folks who introduced us to CKS and Mark [Kvamme, chairman of CKS], who was part owned by Interpublic at the time, and Hambrecht & Quist did the deal. The whole process took a few months, and we were acquired. It was kind of the first, or one of the first, of the Alley businesses to be acquired. And I think it was the third issue of *Silicon Alley Reporter* and Seth was on the cover, and there was a scatological interview. Seth, you know, foul-mouthed.

*Goddamned* this, *shit, fuck,* whatever, about the business. People accused us of selling out; it got a lot of controversy, and raised the hackles of a lot of firms. Little did we know that everybody would be selling out soon enough.

**SETH GOLDSTEIN**   We sold SiteSpecific for stock, so it wasn't a fixed price. It was roughly eight or ten million [dollars], which at the time was an enormous amount of money. I just really felt it was time. We were kind of running on fumes. We had been innovating a number of business models, and we realized that unless we had capital, we weren't going to be able to grow to keep up with the business. We wanted some more infrastructure, and we wanted a strong partner that would allow us to compete against everyone in New York who had raised money or gotten bigger.

I think it was one of the first deals done in the Alley where an agency sold itself, and it was great. But it was sad as we saw what we created get dismantled. They didn't keep the brand, and we all learned a lot in the process. Everyone got out in '97, early enough to get involved with a next-generation Internet company.

*Were the other agencies structured to facilitate massive growth successfully?*

**SETH GOLDSTEIN**   No. Fundamentally, it's a body-shop business. You can talk about your core technology and how it makes you more efficient, but as you can see right now, everybody's laying off people. Agency.com's stock is down to 3 [1.5 by March 2001]. Anyone that lasts until next fall will probably survive, but along the way there's going to be a lot of failures.

**KYLE SHANNON**   At some point I had to let go of UrbanDesires; my wife took that over. A year and a half ago I had to let go of WWWAC, because I wasn't doing it a service and it was distracting me from Agency. So I'm now fully dedicated to Agency, but those two things still exist and are still out there making people happy.

**ALAN MECKLER** | FOUNDER INTERNET WORLD, INTERNET.COM |
There are a lot of business models that don't work, like the whole Inter-

net consulting thing. There are a lot of companies in New York and across the country, but consulting is consulting is consulting, and to have the market valuations that some of these companies had in the advertising and consulting space was nothing short of ridiculous.

**STACY HORN** Kyle Shannon, when I met him, was an out-of-work actor. So for him to hit that kind of success was amazing, and in fact, I use him as an example every year with my students at ITP [Interactive Telecommunications Program at NYU]. You can have absolutely no background and succeed like Kyle.

**DOUGLAS RUSHKOFF** I admire Razorfish. I'm sorry for the drastic loss in stock value they're going through right now. But I admire them, because what they tried to do was maintain the original balance in Silicon Alley culture between smirky, snot-nosed brats who understand the technology, and bumbling, dithering, overweight, middle-aged, suited fools who don't understand, and who depend on us to tell them what's going on. The culture of Razorfish is, "We get it, and you guys don't." That's the sort of fun, adolescent countercultural little zing that I got off on in the early Internet. I think Razorfish's clients still feel like, These kids kind of get it, and we sort of don't, and we need them. And through it all—even now, though he's behaving in a way that's responsible to his investors because he has to—Craig has still maintained some of that Mick Jaggery, rock 'n' roll vibe.

# THE RISING TIDE

*In a brand-new industry whose very infrastructure was continually evolving, traditional rules of competition no longer seemed to apply. In the Alley, the concept of "co-opitition," or companies competing in a marketplace while supporting other companies and the industry at large, was extremely popular, especially in the first few years.*

*As quickly as entrepreneurs began joining together with newfound partners in apartments and offices late at night, calculating their risks and shoring up their resources before starting up their new ventures, organizers began to gather together colleagues into industry foundations like the New York New Media Association (NYNMA) and the World Wide Web Artists Consortium (WWWAC).*

*NYNMA was the brainchild of Mark Stahlman (the investment banker who helped take AOL public in 1992) and Brian Horey, a venture capitalist. Out of his evening salons Stahlman joined the first industry services people, public relations agents, lawyers, and entrepreneurs, and began to hold events and other efforts to serve the new industry.*

*WWWAC, the World Wide Web Artists Consortium, was formed by Kyle*

*Shannon in 1994 as a meeting place for Web designers and developers. It was a notable testament to his collaborative spirit that he started WWWAC within a few weeks of cofounding Agency.com with Chan Suh. At their monthly get-togethers, WWWACers scrutinized the latest developments in the Web, sharing information with new start-ups, and set up an email list for members to use in soliciting help from colleagues.*

*Perceiving an opportunity to mobilize Internet workers for the sake of education, nightclub owner and dot-com entrepreneur Andrew Rasiej started MOUSE (Making Opportunities for Upgrading Schools & Education), an ambitious effort to wire public schools in New York City.*

*There is no question that the efforts to build community succeeded. Together, and with the help of a starmaker named Jason McCabe Calacanis, these organizations nurtured a well-oiled business machine that from early 1998 until March 2000 spit out a few dozen IPOs and facilitated the funding of more than eight thousand start-ups. After 1998 New York City was a technology city running at full speed, and thousands of companies, with hundreds of thousands of new information workers manning their engines, emerged from the ether to compete in the great new game.*

*But back in the early 1990s, when the country was beginning to learn what happened when you stuck a telephone cord in the back of a computer, a dial-up community named Echo brought together early adopters of the new medium. Echo was an innovative online hangout started in 1990 by Stacy Horn, a disenchanted corporate fugitive who began the project while a student at ITP, the Interactive Telecommunications Program at New York University. Modeled after San Francisco's The Well, Echo was an online home away from home for writers, artists, and intellectuals.*

**STACY HORN**   At the time I started graduate school at ITP [in 1986] I was really, really scared. I was in my early thirties, but I had been out of school for a long time. And I was really intimidated, because it's graduate school, and you're surrounded by people who "think," and I really didn't see myself that way. I remember when I started people would raise their hands and ask these really intelligent questions and I'd be sitting there just trying to think of a smart question to ask. I could never think of anything, so I never said anything.

Red Burns was teaching this class called Applications of Interactive Telecommunication Systems. Every week we had a different speaker, and we had to write a paper on something the speaker had said. I started a half dozen papers, but I just couldn't finish anything I started and I just hated them all. Finally I said Fuck it, and I wrote a play called *Corpse in Space*. One speaker had talked about implementation, and so I thought, I'll write about starting a business—except this is a business where the world has run out of space to bury people so they shoot them out into space into orbit with Earth. So you could stand in a certain place and know your loved one would go by. It was just something I had fun with.

The day I had to hand it in I was terrified. I was sure this was not going to be acceptable. When everyone was handing in their papers, I put mine in the pile and ran out. The next week I get there and Red Burns is at the end of the hall and she starts screaming my name. I kept thinking, "Oh God, oh God." But she came up and hugged me and kissed me, and said all she'd gotten was boring paper after boring paper, until mine. That completely set the tone for me and I spent my years there doing whatever I wanted, and she supported me. She can be difficult, but as far as I'm concerned she's a saint.

**MARISA BOWE**  I grew up in Minneapolis. My father worked for Control Data Corp. In the early 1970s, Control Data bought this computer network called PLATO, and decided to develop it for education. It was the first user-friendly computer service with real word commands instead of obscure programming. It was really radical for its time. My father was a PR executive there; he had nothing to do with the computers himself. But the head of the company wanted all the executives to understand the product, so we had a terminal in the basement in 1975 or 1976.

I had no interest in programming, but somehow I found out that you could talk to people on this network. It actually predated the ARPANET, which was being developed right around the same time. I started hanging out and chatting, and it was all guys and me. I thought this was wonderful. I was sixteen or seventeen and kind of shy, but I was getting all this attention.

I found out much later that this was the first social environment online ever. There was a real community of people, but I just chatted with people one-on-one. I would find people randomly—like my dad's boss's son. I just got obsessed with it and loved it. I remember thinking, This might be good, because computers might be big in the future.

After a couple of years of talking on PLATO, I forgot all about it for another decade. Then I started reading about The Well in a Whole Earth Catalog or something. And I thought, That's like PLATO, but with interesting people involved instead of just programmers. So I tried The Well, but I hated it because I hate that West Coast, Deadhead, Mondo 2000 crap. It was just such a different sensibility. I mean, I'm just old enough to have experienced hippies the first time around, and I didn't really like it the first time around. It's sort of toxic. Also, you had to make a direct-dial call to San Francisco, so it was expensive.

Then there was an article about the EFF [Electronic Frontier Foundation, a hackers' rights organization] in the *New York Times Magazine* that mentioned Echo. So I logged on and that was it. Four years later I looked up from my desk.

**STACY HORN**   Echo was instantly a very odd, eclectic, dysfunctional little community, and it still is. I lucked out because before I was even open, this woman from the *New York Times* wrote an article about us, and it was like she was my PR person. The article was called "Coming to the East Coast: An Electronic Salon," which is exactly how I wanted people to think of it. And even though we weren't open yet, she described us as this place for artists and writers, as this very interesting Gertrude Stein type place. So I got all these phone calls from playwrights and actors and musicians—exactly the kind of people I wanted—and those were my first users. When computer people came online and saw that we were talking about opera and not games, they left, which is also what I wanted.

**MARISA BOWE**   It's really abnormal that someone my age would be into this stuff, and it was all because of this fluky thing with Control Data. I had worked in this underground alternative television network, this lefty TV version of an underground newspaper, and I had always been interested in personal technology for communication around the mainstream, so it was sort of a natural thing for me to get into.

Back then Echo was much more female than any other online space, but it was still pretty geeky. People were having very intelligent conversations, but also blaming each other for having romances. It was the same type of Peyton Place that Echo has always been, just on a smaller scale. I had been really into 'zine culture, and this to me was like a living, breathing, ongoing 'zine. Here were interesting, underemployed, smart people like me, playing these weird games and talking about stuff, and it was just fascinating to me. I remember sitting up night after night, laughing my head off. The capacity for projection was endless, so I got a lot of attention for being funny and smart and I just lapped it up. I think that's a big part of what the online dynamic is. People are starving for attention; if they're not that successful, online they have another chance. It's like high school all over again. I got to be prom queen, so that was like cocaine for me, like *Praise me more!*

**STACY HORN**　At the time, the only people online were government people, business people, and students—and only students studying computer science or science. That was true until 1993. I taught this course at ITP after I graduated—which I still teach—called Introduction to the Internet. I would teach students how to go online and move around. Every semester they would start out so excited to get on the Internet, and then when they got there they were like, "Who cares?" These were students more interested in the arts, and it was all computer students talking about the latest game they'd downloaded or the latest program they wrote, or scientists talking on a level no one could follow, or business people, and that was it. There were no websites. They'd get online and go, "This is the Internet? So what?"

　　When I started Echo I had a hard time getting people to finance me, because I had a hard time getting people to believe the Internet was going to be big. Visually it was ugly, so even though I thought anybody with half a brain could see that this was going to change and evolve, I wasn't shocked that I couldn't convince anybody.

**CLAY SHIRKY**　An Internet community either has a center or it has an edge, but it has to have one or the other to be a community. Stacy's community, Echo, was an edge-oriented community. Which meant that when you paid your money you were an Echoid, and if you didn't, you weren't. And within Echo there was this whole internecine world of who was sleeping with whom and who was mad at whom. But paying your money was enough to get you in, whereas on Usenet [a public bulletin board system] anybody could get there, so to be a member of the community you had to gravitate toward the center, which was the Old Hats [veteran users] list. It was really interesting, the way the technology affects the culture.

**ALICE RODD O'ROURKE** | EXECUTIVE DIRECTOR, NYNMA | NYNMA was founded in the summer of 1994 by Mark Stahlman and Brian Horey. Brian was also one of the first VCs investing in the Alley, with a group called Lawrence, Smith and Horey. They and other like-minded people were tired of flying to San Francisco and running into other attorneys and accountants in the airport. And they said, "We should do this in New York one of these days."

**BRIAN HOREY** | COFOUNDER, NYNMA | NYNMA formed because this new media thing was starting to happen, and we were concerned that New York wouldn't play the role in it that it should. At the time we got it started, CD-ROM development was still the leading edge of the new media industry, and we thought too much of it was going to the West Coast. We needed more infrastructure and a sense of community if we were going to build something here. Things like the browser hadn't really been popularized yet, but you could see the growth in traffic on the Internet, particularly in terms of email. There were also an awful lot of people applying digital technologies to traditional creative and artistic processes, but they were spread out in lots of different businesses—advertising, graphic design, music, publishing—and they really didn't have a place to coalesce and find each other. You could also start to see that this could become an interesting economic and technical phenomenon, and that there was going to be a way to make money, though it wasn't clear at the time how.

I proposed to Mark Stahlman over lunch in the summer of '93 that we start something to try to coalesce this community, which was here but was hard to find. So we brought together a group of people, probably thirty or forty, in the spring or early summer of '94, had a cocktail party in Mark's loft, and put forth this idea. We asked them to help us do it, and a lot of those people became charter members of the association. Some of them had come from a Cyber Salon series of dinners that Mark had been having for a year or two before then, but not all of them.

**ALICE RODD O'ROURKE**   I was the head of the three-hundred-person economic development department [Empire State Development] for New York State, and in 1996 NYNMA released a survey that said that the industry would create between forty and a hundred and twenty thousand jobs in the next three to five years. Being a bright button in the economic development world, I said, "Gee, New York is down four hundred and fifty thousand jobs in 1996. Maybe we should be paying attention to this industry, so that we can close that gap." Having done business development and venture work, I knew that even with half the number of jobs and double the projected time, there still wasn't anything else that was happening in New York that could create that many jobs.

New York was an exporter of jobs; it was one of two states whose population was declining despite great immigration. A year and a half later, when I was looking to leave the state, I saw that NYNMA was looking for an executive director and raised my hand and the rest was history.

**MARK STAHLMAN**   The NYNMA CyberSuds events were really the community-building mechanism, more than anything else in this town. When we finally got—I think it was close to three thousand—people in the lobby of the IBM building uptown, then we knew that something had actually happened. Some people might cite the Cyber-Suds they had at the Playboy offices, too. That was almost impossible to do, because it was on the top floor of a building and only had a fire capacity of three hundred fifty people, so when there were five hundred people on the street waiting for the elevator, that's when it dawned on some other people that something had really happened here.

**SETH GOLDSTEIN**   The early CyberSuds were very important. I met one of my first clients at an early CyberSuds. They provided a forum for people to meet with each other. They were in the right place at the right time. I don't think they single-handedly stimulated Silicon Alley, but I think they provided a venue for people of like minds to get together and talk.

**BRIAN HOREY**  We wanted people to start the process of networking and collaborating. One of the things I had seen in the software and PC businesses in places like Boston and California was that no single company ever had all the assets and skills and technology it needed to be successful in its business. It had to collaborate with other firms and other people, and we thought things would be the same here. Part of what we wanted to do was just get people in touch with each other so they could be more effective in their businesses.

We started out with maybe fifteen or twenty people at the first CyberSuds, and by the fall we had more than a hundred people at each event. We asked people to become members, and then we hired an executive director in December or early the next year and started to have events beyond CyberSuds.

**ALICE RODD O'ROURKE**  If you look at the history of what NYNMA has done, it very much reflects the late history of the development of the industry. The very first activities were networking activities—bringing people together on any given month to go to a CyberSuds. We did this industry survey, and let people working in their living rooms—or in someone's office that they managed to talk themselves into for free for six months—know that they're not alone, there are thirty-five thousand of them.

That was followed up fairly rapidly, probably because of the founders' financial backgrounds, with the realization that if they didn't get money, they couldn't get big quickly enough to register. So we developed the Venture Downtown program and followed up some time later with an Angel Investors program. Some of the earliest new media successes went through the NYNMA Venture Downtown program, including iVillage, Yoyodyne, EarthWeb, Agency, Kozmo, and Register.

Then the companies needed a place to get exposure. So we created Super CyberSuds, which is a tabletop trade show, now in its sixth year. There aren't many institutions in the Alley that are older than Cyber-Suds and Super CyberSuds.

**OMAR WASOW**  Eventually NYNMA started having a lot more people in suits and a lot more people who weren't necessarily practitioners, in the sense that they didn't get their hands dirty with code, weren't deeply into the Internet. They were more the nontechnical folks who were getting involved in the industry, and that's wonderful, but it made you realize it was no longer this elite club anymore. The party's been crashed. For the most part that's been a wonderful thing, but when you're this early arrival, there's this certain nostalgia you have—like, Why did everyone else need to get access to fire? We were having a lot of fun with it by ourselves.

**CLAY SHIRKY** There was an amazing time in the early days on Panix [Public Access Networks Corporation, the oldest commercial ISP in New York] where when someone sent you an email you had to answer it. One, because you only got ten pieces of email a day. But two, because they were a peer. The only people who were on the Internet were either hard-core engineers, or weird people like yourself who had somehow found their way into this magic kingdom.

As the network became more normal that feeling went away, so if you ran into someone else on the network, they didn't feel like a peer anymore, there were so many people there. And there was a period on Usenet that was referred to as "the long September." Every September, famously, a new group of college students would get access to Usenet, and come on and act clueless and stupid. And people would flame them until they either left or decided to become part of the community. But starting in '94 or '95, September never stopped.

**JASON MCCABE CALACANIS** I started sending people information about parties, and I realized that email was the most powerful thing in the world for publishing, just clearly, and I said I have to have all the email addresses of everybody in Silicon Alley. And that was one of the things that NYNMA had over me. New York New Media Association had two thousand members and all their email addresses. I really looked up to NYNMA, and really wanted to be involved in it. I probably made one of the best mistakes of my life when I went to the NYNMA board meeting after I'd written [in *Silicon Alley Reporter*'s premiere issue] that NYNMA was out of touch. They invited me to come to Mark Stahlman's loft and explain to them why. They brought myself, Stefanie Syman, and Craig Kanarick there and said, "What do you think we should do to change NYNMA?"

I said, "You should change the mandate of NYNMA. The mandate for NYNMA should be for the entrepreneur and the freelancer." And so they changed their mandate to that. Then I said, "You should start an angel network where you match angel investors with people," and they started that.

And then I said, "You should put people who are actually entrepreneurs and in Silicon Alley on the board, like myself, Craig, and Stefanie Syman." Two months later they sent the email out that they'd put Stefanie on the board, Craig on the board . . . and they snubbed me. So I emailed them to ask them what happened, and they wouldn't tell me. They never put the board up to vote, and they snubbed me.

And that was it. At that point I said, you know, The world's against me, everybody wants to try and stop me. NYNMA doesn't want me involved—*whatever*. I was hell-bent on beating NYNMA and beating @NY, and that was my career. I was going to win just like Jordan was going to beat the Knicks and whoever stood in his way. That was my mission. I'm not going to lay down and take it. They stole my ideas, and I said, That's great, you can steal those three ideas, but I have a lot more ideas than you do and a lot more time to execute them.

So I put myself out there. NYNMA had a print magazine, too, in the beginning. They stopped doing it. They did events, but then I started doing events, and it was very clear that I could do them better than NYNMA.

Competition was the motivator for me. I was twenty-five years old, and competition is a very immature motivator, but it is very powerful—and I didn't like to lose.

**CLAY SHIRKY**   We started to have a sense of ourselves as a community. There started to be tropes of dressing and so forth: all of a sudden you could walk into a coffee shop and think, That guy, he works in the industry.

We would always hang out at Eureka Joe's, a coffee shop on Broadway near Twenty-third Street. That was a sort of official commissary. Eureka Joe's was hilarious, because up until that point it had been a place for fashion models and their agents. These waves of people would come in, and you could just sit there and point: fashion, Internet, fashion, Internet.

There were all those hot spots. Even that stupid diner on Nineteenth and Broadway, that teal blue diner with the big glass windows, became a deal-making hangout for Silicon Alley. I used to recruit people at the Barnes & Noble at Sixth and Twenty-first. If I saw someone reading an advanced PhotoShop book, I'd say, "Excuse me, are you looking for a job?" Sort of the financial equivalent of cruising.

It's still a personality-driven business where there's a couple hundred people and you've got their email addresses and they've got yours. But now you go to these parties and you think, If these people work in the Web, maybe *I* don't. I don't see a soul I recognize here. And then you run into someone from the old days and you both get this wild-eyed look, like, Oh my God, what happened?

**KEVIN O'CONNOR**   The first people that I really remember as being part of the Alley were Seth Goldstein and Scott Heiferman. We were trying to have them as customers. If it was a community, it was a pretty loose community. Then we started getting involved with NYNMA, and it shocked me. I didn't realize there were that many people in the community—I had thought that it was only a couple hundred people.

**SCOTT HEIFERMAN**   There was a community, but the idea that it could be labeled, that it was anything of real importance, was kind of funny to me. The community formed because this was important, because the Internet was a revolution, it was the Resistance, it was out to do big things. But at the time the idea that the community itself could be labeled was funny.

**JAIME LEVY** | FOUNDER, ELECTRONIC HOLLYWOOD | There was the WWWAC group, the World Wide Web Artists Consortium. I went to the first few meetings, and there were maybe twenty of us. I remember meeting Kyle Shannon back then, and Chan, and Scott Heiferman, who turned out to be my third cousin. Clay Shirky and Howard Greenstein, too.

At that time we would have a projector, and at one meeting Kyle showed— Guess what came out this week? Tables. He showed a demo of tables, and everyone said, "Oh my God, this is going to change everything." And I went back to Word.com and I'd say, "We've got to do an artistic piece with tables in it." Every week there was some new HTML code to exploit.

**SCOTT HEIFERMAN**   When I was at Sony in 1994 I started going to meetings of the WWWAC. It was absolutely enthralling. It was so amazing to me, having just been from Iowa, where you could count on one hand the amount of people who had any interest in the Internet, and even they were coming from a technology angle rather than a cultural angle.

I remember going out for pizza after WWWAC meetings, which were maybe twenty-odd people. It really was, at least from my perspective, an innocent thing. The word IPO hadn't been applied to this world yet. Sure, when you start a company, obviously you're thinking about money, but there was no way that anyone could start a company in this industry after 1995 and have it be as pure and innocent from a money perspective.

I started my company without the expectation that Internet companies would be really valuable, or that it would be a road to riches. It was damn interesting, and I can't speak for the other people involved at the time, but for the most part, it was about sitting around and saying, What the hell will America look like when half of it is on the Web? What will happen when most people are using email?

Obviously things that are true today were not true then. That's what was fascinating. It wasn't about business directly; it was about the role we could play in it, and what new things could be invented and how will the world shake out. It wasn't a boardroom discussion. It was over pizza.

**KYLE SHANNON**   After I met Chan Suh I realized that there were people in the world that knew more about this than I did. But it was so new that I probably knew some stuff that other people didn't know. WWWAC was initially going to be by invitation only, for people who were actually doing something on the Web, not just people who wanted to. You actually had to have some knowledge. I'd say, "Let's drop whatever our motives are here and just ask, 'How do you do that?' "

On a selfish level, I wanted to surround myself with people who knew more than I did. And it worked. Howard Greenstein was one of the original members. Peter Seidler [chairman of Avalanche Solutions, which merged with Razorfish in 1998]. Chan Suh. What ended up happening was I realized that having it be a closed group was fine, but not really where it was at, and that we should open this up to the community.

In those days it was not at all about connecting to trade business cards to get venture capital. There was no venture capital. This was about showing up, sitting around the table, and talking about what happened this week. "Oh, there's a new version of the browser? Can someone show that?"

There was a real camaraderie of invention. It felt like what I imagine the early days of the '60s protest movement must have felt like—all that hope. We can change the world, for everyone's sake. There were no divisions yet, and no one had really made a voice for themselves. There were personalities and there were ideas, but it was really free-flowing.

Very early on, we got senior-level people from Microsoft, AOL, and Prodigy to come in and say what they were up to. And if we didn't like what they had to say, we would just blast them for half an hour. We would just yell at them. People knew not to put any vaporware in front of that group, because they knew we'd find them out and chop them down. That was kind of fun, and it gave the New York community a very strong sense of authenticity and integrity.

Every week we would go around and introduce everyone in the group, and you started to see people that were there all the time, and some new faces, and then people you hadn't seen before. There were always ten or fifteen thinking about quitting their non-Internet jobs. It became like a support group for people wanting to quit their day jobs. In the early days WWWAC was really about encouraging people out of their daily lives and into whatever this was going to be.

**SCOTT HEIFERMAN**   I really do believe that Kyle Shannon, in the infant stage of his own company, started the WWWAC as an organization that was about sharing because he thought, This isn't about Agency.com hoarding everything, it's about how we're going to rise and fall together. Just like the Internet is better when there are more sites to surf on, an open environment is better than a closed environment; an open infrastructure is better than a closed infrastructure; and that happens to translate to the way a lot of business has been done. I think there's a willingness to help others in the business that stems from the open infrastructure of the Internet.

**ANDREW RASIEJ** | COFOUNDER, DIGITAL CLUB NETWORK | The business community got together and started working on some outreach mentoring programs and other kinds of things at Washington Irving High School, down the block from Irving Plaza, and they invited me to join them. So I went over and I saw that kids were typing on typewriters. And I thought it was somewhat ironic that we were the most technologically advanced culture in the world and yet these kids were using machines that were fifteen or twenty years old.

The school had one computer lab, using old 486 machines, but we figured out that we could get them to work on the Internet if we hooked them up properly. I ordered a T1 and sent an email out to ten of my friends saying, "Will you join me on a Saturday to wire this school to the Internet?" I got a pretty immediate response from people that said, "Hey, sure, what would you like us to do?" I didn't know, actually, so I wrote back saying, "Why don't you forward my email to ten of your friends and by the time everybody mails it back, I'll have figured it out." We had a hundred and fifty people show up on that Saturday. I never picked up the phone. It was all done by email.

There was a lot of enthusiasm around the idea, despite the fact that I still didn't know what we were doing. But that was how MOUSE was born. I remember holding a ladder for Gene DeRose from Jupiter Communications, and I realized that Gene had a whole bunch of skills he could use to help with education technology, none of which he was using that day. Instead he was up on a ladder, trying to pull some cable. It occurred to me that we could build a database, allow people to register as volunteers, identify their specific skill sets, and then deploy them in the school system based on need.

Since we launched MOUSE almost three years ago, we have over two thousand people in our database, we've wired about fifty-five schools, put about seventy-five thousand kids online, and have developed all kinds of programs like designing online school newspapers, internship programs, and mentoring programs, where all these people in the industry can do something philanthropic.

**ALICE RODD O'ROURKE**   Getting schools wired is obviously a step toward the adoption and understanding of the Internet. If the Internet isn't there, then even with extraordinary teachers these extraordinary students are not going to have much experience with it.

In the early days of our internship program, and in the early days of MOUSE, we collaborated to try to make one plus one equal more than two. For instance, when MOUSE wired Martin Luther King High School, we made sure it also became part of our internship program. You need to show students that there's something big you can do with this, and it's important. It's a career, a lifelong kind of thing.

*What happened to the original members of the community?*

**SETH GOLDSTEIN**   Of the early people, a quarter of them sold out, got serious and tried to make a lot of money; a quarter of them became upper-level management; a quarter of them toiled away in obscurity; and a quarter of them became freaks.

# WHEN CONTENT WAS KING

*In 1995 conventional wisdom in the Alley held that while California would always lead the development of Internet infrastructure, the network would be worthless without rich content to fill its pipes. The aphorism "content is king" was so common in New York that its validity was scarcely questioned, and it wasn't hard for anyone to figure out that if content would be king, New York City would be its throne.*

*Early content pioneers thrilled to the idea of a new online magazine format in which distribution and printing costs would be nonexistent. At first the idea was that by selling subscriptions, online magazine publishers would avoid the overhead costs paper magazines endure, and reap a greater profit.*

*But it quickly became clear that subscription models for unknown proper-ties on the Internet wouldn't work, and content proprietors were forced to turn to advertising for revenue. Looking at the accelerating numbers of Web users, they felt confident they could turn a profit through advertising. A magazine format based on ad revenue seemed like a golden ticket to riches, the banner ad the new cash cow.*

*What they didn't predict was that CPM (cost per thousand ad impressions) rates would quickly drop, and it would become increasingly difficult if not impossible to support most of the developing sites. Very few online magazines and content sites were commercially successful, but the truth is magazines have always been a fragile business, and few actually make significant money.*

*But along the way, fundamental changes were being made in the distribution of information and art—changes that, for a time, investors were clamoring to underwrite.*

**ESTHER DYSON** | CHAIRMAN, EDVENTURE | Content began to become more important. Originally it was a high-tech business, PCs and programming that came from California. But when people realized, "Oh, we actually need content—this is an advertising medium," suddenly they started to think about New York again.

**KYLE SHANNON** The epiphany I had about the Web came when I was looking at a flyer for a friend's band. I looked at the Web, then at the flyer, and I thought, Shit, this could be online, and they wouldn't have to mail them out, and more people could get to it. That's cool. So then I came up with this idea for a company that would have been a record label for unsigned bands. I had someone who knew a little about money and spreadsheets, and we created a little financial model, and he said, "Well, to break even you're going to need relationships with a hundred and fifty bands." All I had were a couple of musician friends, and they were freaks, so that wasn't going to work. Then we flipped the model around and thought, If you did a magazine, getting content from people and repurposing it, then at least you could break even without all that overhead. But what kind of magazine do we do? And of course I started out with a sex magazine because, well . . . it's sex. Porn will win. But eventually my wife and I got *ooged*-out about that and we shifted the editorial view and decided to call this thing UrbanDesires. It still had that sexual undertone to it, but the idea was that in the city so *much* is desirable. Food is desirable, and movies are desirable, and books and toys and this and that. So we created this culture magazine that explored the darker side of all these things. And we ran that for three and a half years.

I created UrbanDesires in August and September of '94, and I had it on my local machine but I needed to get a server to be able to put it up.

I put it up and none of my pictures showed up, and I'm like, "Fuck." I was on Echo and very active in that community at the time. They had been talking about the Internet for about a year, and about the Web; I started the Web conference in Echo. So I had a lot of connections in there. One of the regulars was this guy named Vibemail; that turned out to be Chan Suh, who's now my business partner. He was doing Vibe Online, so when my pictures didn't show up I said, "Hey, Vibemail, my pictures didn't show up, what do I do?"

And he said, "Are you on a Mac?"

I said, "Yeah."

And he says, "Oh, it's going to take too long to explain. Why don't you just come down here?"

We met and he gave me some technical advice, and then we launched UrbanDesires in earnest. I was helping him with some design stuff, and we just started talking about why Myst was a great game and what Myst and the Web had in common, and wondering if we could do work as compelling as Myst in an environment that's as constrained as the Web. Back then it was all gray backgrounds, no center tags, just text and maybe a picture. Not much you could do.

The moment when the power of the Internet hit me square between the eyes came we launched UrbanDesires in November '94. About a month after I got an email from someone on Echo, a French person, saying, "Hey, they just did a full-page article on UrbanDesires in this French newspaper called *Liberatione* in Paris."

I went to one of the international newsstands on Forty-second Street and bought the paper, and there's UrbanDesires. There was a full-page article on something that only a month ago was nothing more than a couple files on my computer. If you think about the effort that goes into putting on a play that's going to run on Theater Row, and you get sixty people total to come see it over two weeks, you've done good. I did a similar sort of thing, putting together some content and sticking it up just like you stick up a play, and all of a sudden people in Paris are reading it and writing about it. In a *month*.

*A major turning point occurred when AOL and other online services switched to a flat-rate pricing system. Many content sites had made most of their initial revenues from fees paid by online services as a percentage of billable hours*

*spent by users on their site. Adult chat was one of the primary revenue gener-*
*ators for online services. But anything that kept a user from hanging up the*
*modem was valuable and made money.*

*When online services switched to a flat-rate plan, content sites had to start*
*relying on online advertising alone, and they were never able to catch up with*
*their growing costs. iVillage and other aggressive content/community plays*
*made some money early on selling sponsorships, but eventually the rules of the*
*market would completely dilute the revenue per site of online advertising. The*
*more successful any Web site seemed to be, the more well-funded competitors*
*would emerge to threaten their market share.*

**JASON CHERVOKAS**   In the era before "all you can eat" Internet
services, where all Internet services were done on a per-minute basis,
the business models were very different. The reality was, as long as
AOL was charging per minute, and then parceling money back out to
content providers based on time spent inside their content sites on
AOL, it was a very different business model. Ultimately, though, AOL
realized, "Content really isn't our business; connecting people with one
another is our business."

**NICHOLAS BUTTERWORTH**   It became clear that SonicNet
needed to find a partner. I'd become friends with Becky Savell, who ran
the music channel for Prodigy, and talked to [Prodigy CEO] Ed Bennett.
Ed called to say he wanted to talk about possibly investing in SonicNet.
Prodigy had incubated their own content thing, called Stim, with Steve
Raymond and Mikki Halpin. That was another early content play.

We sold 51 percent of our company to Prodigy in 1995, for a promise
to invest a couple million dollars going forward. Prodigy was intense.
Ed Bennett basically inherited this company, which had been a joint
venture between IBM and Sears, that had the reputation for being a
dumping ground for failed IBM executives, and had spent huge
amounts of money on equipment and mainframe computers. They had
a technology group in Yorktown, New York, had their headquarters in
White Plains, but Ed felt IBM was too stodgy to be a media play, and
they had to be in New York. So we opened this cool downtown office at
632 Broadway. I prided myself that in the whole year and a half that we
had Prodigy as an owner, not once did I ever go to White Plains.

What happened with Prodigy was that Ed hired a lot of young, smart MBAs from the media business who didn't know much about online services or software. They interfaced with a bunch of old-style engineers in Yorktown, who didn't know a thing about consumers or media. I always felt that what hobbled Ed—I thought his vision was correct, but he failed to see that your ability to develop a compelling user experience was always either limited or enabled by technology.

And Prodigy's technology was terrible, from a user-friendliness standpoint. Where Steve Case [at AOL] was focused like a laser on usability and communications, Prodigy had an email system that was so unfriendly to users that their own executives didn't use it to communicate with each other. This was a cardinal rule at Microsoft—use your own product. Not at Prodigy.

At any rate, SonicNet was supposed to be the first of a number of venture investments by Prodigy in the content media space, and instead we ended up one of two that got completed before IBM and Sears pulled the plug in the spring of '96.

We really got there just as they got going. Or, rather, IBM and Sears took away the punch bowl just as the party got going. By the time the partners decided to quit their funding, Prodigy had already burned through a billion dollars. If you think about it, for a billion dollars you could have bought every single website times ten. You could have owned Yahoo, Amazon, Infoseek, AOL, and Excite. It's a staggering figure. From then on, Prodigy didn't really have a lot of money for us. We were far from any kind of profitable business model. We went out hunting for money again in the summer of '96.

*FEED (feedmag.com) was a pioneering online magazine that took an intellectual approach to popular culture and technology.*

**STEVEN JOHNSON** | COFOUNDER, FEED | The industry was really undifferentiated. You had a bunch of companies that didn't really know what business they were in—and didn't really *have* to know at the time, really. So Agency was publishing UrbanDesires, and they were very serious about it; it was a real product, which they updated way more often than we did. And yet they were also doing corporate

Web development. Kyle and Chan may have an alternative view on this, but from the outside, it was almost as though corporate Web development was subsidizing this fun publishing venture, and it wasn't clear in the very early days which would dominate.

Nicholas [Butterworth] had gotten a company called the New York Web to agree to host this political site for free on their servers, which was kind of a big deal back then. So that night Nicholas got me to agree to look into this stuff, and we ended up having this epic argument about the Web versus BBSs that went on until about four-thirty in the morning. I remember the argument very distinctly, despite the amount of alcohol that had been consumed.

It was back in the day when there were no manuals for HTML, so I downloaded pages and looked at how they did the code, which is not all that hard to do but seemed pretty impressive at the time. So I built this little site called Mpower and I took it down to the New York Web, and they liked it and asked me what else I did. I told them I was still at Columbia [in the Ph.D. program], but I was a writer and had written for *The Guardian* and *Lingua Franca*. "Oh, you're a writer. We want to start an online magazine, and would you like to be editor in chief, because we don't know any writers." Classic early moment. They just wanted to start a magazine because they could.

They had an idea for what basically became Total New York, a *Village Voice* kind of thing online. I said no, but maybe I'd like to edit something else. So I went home and wrote up this little manifesto for FEED. I sent it to, like, ten of my friends, all of whom I thought would immediately jump on board, and to Stefanie, whom I'd just met. Everybody else was like, "That's a great idea, Steven, but you'll never do it." But Stefanie, fortunately, didn't know me well enough, so she said, "That's great—let's do it," and jumped in immediately.

We started working out of the New York Web's office in March of '95, and launched on May 17, 1995.

**STEFANIE SYMAN** The New York Web was going to be the business infrastructure of FEED. They were going to sell our advertising, provide office space, reception; we didn't even know, really, what we would need, but anything that was "business" they were going to take care of.

FEED at the time was literally one little desk, two broken chairs, and a computer.

**STEVEN JOHNSON**  One PowerMac 7100, split between the two of us.

**STEFANIE SYMAN**  And Steven and I would sit behind the desk, on the same side, sharing a phone, using the computer at the same time.

**STEVEN JOHNSON**  Like, "Okay, you want the keyboard? Why don't you drive?"

When Voyager arrived from L.A. in 1992–93 it was kind of an epic founding moment for Silicon Alley. They'd been based on the West Coast, but they decided to move to New York because they were in the content business, and content lived in New York, which really became the mantra of Silicon Alley. Both Stefanie and I had friends who worked at Voyager. Bob [Stein] was the first name we got involved with FEED. We had raised about fifty thousand dollars from extended family, along with some money I had, and we wanted to put together this dialog in electronic text. And we figured we probably had to pay people a thousand dollars to contribute to this little round table.

So we went in to see Bob Stein. We were these two people starting a web 'zine at a time when nobody was doing a web 'zine, and we're doing this kind of academic roundtable discussion on electronic text. And we said, "We'll pay you a thousand dollars to participate."

He was, like, *"What?* Can I just ask, where are you getting a thousand dollars? Sure, okay, I'll take it." Though he clearly would have done it for free.

**STEFANIE SYMAN**  The corporate structure of the New York Web consisted of one guy named Kris [Graham] and another guy named Stephan [Antonovic]. Kris was the CEO and Stephan was president. Stephan had a design background; Kris had more of a techie background.

One Friday afternoon in May, we're on the phone, and we see this group of people enter the office, and it's someone in uniform, a security-guard type, a woman in a suit, and Stephan. And all of a sudden there's a tension, a total change in the atmosphere—as the scene unfolds and it becomes clear that Stephan, the president of the company, is firing the CEO, Kris.

**STEVEN JOHNSON**   And forcing him to leave the premises.

**STEFANIE SYMAN**   And so Steven and I are like, "Oh, fuck." Kris was really upset; this was obviously not a good scene. And we wanted to get out of there pretty quick. I passed Steven a note saying, "Is FEED backed up?" And he doesn't even answer; he starts throwing floppy disks. Within five minutes we had FEED on three different floppy disks. The scene was still unfolding, and we slunk out of the office, knowing we might not even be able to get back into the office. It turned out that Kris had given Stephan a majority of the company, so he did technically have the power to fire him after all.

**STEVEN JOHNSON**   The judge in the case that quickly resulted ordered that the office be taped up. No one could get into it until they'd had this hearing. The servers went down, so FEED went dark ten days after its launch.

We knew we had to find another home, so there was this extraordinary period right after launch where we were walking around the city with these three floppy disks—our little FEED. All we wanted was to take it to its new home.

**STEFANIE SYMAN**   We visited Agency, which at the time was in the Time & Life Building, so Kyle Shannon and Chan Suh were in this little corner of a room. We saw DTI, where we ended up. We talked to Bob Stein at Voyager . . .

**STEVEN JOHNSON**   It was a little tour of the early beginnings of Silicon Alley.

**STEFANIE SYMAN**   And then the funny thing was, we had this woman from *Nikkei*, the Japanese newspaper, coming to do an interview the Tuesday after Memorial Day weekend. We weren't sure, but we had a hunch that we wouldn't be able to get in the office. So I got there early because I'm the early, anal one of the team, and I kind of intercepted her by standing downstairs on the sidewalk. All of a sudden, this big, burly guy appears behind me.

**STEVEN JOHNSON**   I showed up a little late, and Stefanie was talking to this reporter. I can see she's uncomfortable from the way she's talking: "I think we should go eat at Bubbie's, and the office isn't very interesting. Actually, can I just have one second with my partner?"

So I'm like [whispering], "What's going on?"

And she's like, "This guy behind me is a cop. He's guarding the door, and we're not allowed inside."

**STEFANIE SYMAN**   After the legal wrangling let up, we got back into the offices and reconnected with New York Web, such as it was. Then they started demanding their share of our company; we had [originally] said, Sure, have a quarter of the company, and they actually really wanted their share. So we got out of there as fast as possible, because we could see that whatever happened with them, we needed to be in a more stable situation. It was the beginning of our incompetence saving us as a theme throughout our history—that we didn't even have the time or the forethought to get the legal documents written up, or even really pay much attention to the issue at all. There was nothing in writing, and obviously they weren't delivering on anything they promised. So we just walked with FEED.

**ESTHER DYSON**   I met Steven and Stefanie, and I liked them a lot. And, honestly, I invested in FEED because I liked them, and because it should exist, but not expecting to make a profit. You don't say that in public, because it's kind of rude, but the fact is I liked them, and I liked what they were doing.

**STEVEN JOHNSON**   It was very easy to say to investors, here's Salon, which at this point was doing pretty well, and they have forty-five people on staff, and this is where their traffic is and this is where their buzz is; and here's FEED—we have four people on staff. And you have an opportunity to get in very early on this thing. With a little funding, we can grow to six people and really do some amazing things.

Esther invested in FEED because she thought it was good for the Web that we be around.

**ANNA WHEATLEY**  What's difficult for online magazines is that literature is on the one hand very communal, like the coffeehouse scene, but it also tends to be dense and intellectual, and this requires attention, which isn't conducive to reading on the screen. So I never understood a site like FEED. But, you know what, they really plugged and worked hard. They really believed in what they were doing—unlike the community of TheGlobe.com, which sounded like fabrication. At FEED they said, This really is a neat medium, and there really are enough of us out there who like to read and talk about things—we can do something with that. I would like to have seen more success for them, although I do think they will do just fine publishing books and participating in the culture at large.

**CLAY SHIRKY**  Because this is an industry where fifty percent of the population is new in any given year, there is almost no institutional memory. One of my self-appointed functions on WWWAC was to be the institutional memory. At one point Steven Johnson, whom I love working with and whom I write for, got up and said, "Well, we never intended to make a lot of money on this anyway. We were always in it for the culture thing." And I just said, "That's wrong. I was there. You can't trick me." We thought we were going to change everything.

**STEVEN JOHNSON**  All the time while we were launching FEED, we were waiting for Word.com to launch, and we didn't know the people, really, at Word. And we'd gotten that endless banner taunting us on the Netscape What's Cool page, saying "Coming Soon—Word, an online magazine."

**MARISA BOWE**  Word was founded by three people as a business idea: Dan Pelson, who now runs Bolt and Concrete Media; Tom Livaccari; and Carey Earle. From what I understood, they had wanted to start a new magazine in print, but after making the rounds discovered that this was prohibitively expensive and thought, "Well, we should do it online because of the demographic we want to reach." All the media companies sort of laughed them out of the room, or weren't interested, by their account.

Dan knew Scott [Baxter], who was the CEO of Icon [Icon CMT]; they'd worked at Sun Microsystems together. Icon stood for "integration consortium." What they did was, basically, if you had what they called "legacy hardware," they would integrate it with new stuff so you wouldn't have to buy all new computers. That was something they specialized in. Then they thought they should get into the Internet, and like a lot of people they thought, "Well, we'll get into publishing because there will be no distribution costs. We're going to make millions in six months." That's what people really thought.

So they agreed to start Word within Icon. They hired Jaime Levy because she was a celebrity in that world at the time, and they hired Jonathan Van Meter as the original editor because he had left *Vibe* and he was perfect demographically for what they wanted. They had the same idea as Microsoft in hiring a big-name person [Michael Kinsley to run *Slate*] to get a lot of attention in the old media. Jaime had said to Jonathan, "Oh, you should really talk to this woman, Marisa Bowe." At that time the people who knew about online culture were very few and far between.

**CLAY SHIRKY**  Marisa Bowe is the Henry James of the Alley. Marisa is much more attuned to the social feeling or atmosphere in the Alley than I am. I come at it from a much geekier angle. And she was so attuned to Echo in the early days—one of the arbiters of who was hip and who wasn't.

**MARISA BOWE**  I went to interview with Jonathan, and I expected him to be an idiot because I thought all magazine people were idiots and I'd never read *Vibe* and I expected some yucky white guy with dreadlocks or some phony guy. Instead we hit it off instantly. Really hit it off, and adored each other, and we're still really good friends. He hired me to be managing editor, even though I didn't really know what that was. It took me months to get a grip on what an associate editor is and what an assistant editor is. I had no idea what these titles were in the magazine world. People would get pissed at me because I would want to give them a title they felt was beneath them. I just didn't understand it.

Soon Jonathan decided that this project was not for him. I think he had a different idea of the budget and things like that. So he left before the launch, and it was just a classic case of the understudy getting a chance to take the lead role, because they had commitments from advertisers and they didn't just want to back down and shut it down. And I really cared about it. I was just thrilled to be able to work in that area. Jonathan has still never really gotten that interested in the Internet. It was really Jaime and me at first. We had this united vision.

Jaime had done the first multimedia interactive novel, and she had done Electronic Hollywood, her floppy [disk] 'zine, which I thought was just genius. She would go to Mac conventions and diss them in that little 'zine voice. Like those 'zine kids are always saying, "We hate these bands and we hate those bands." She was doing the same thing, but about computers, and on a floppy disk with her own music, and I thought it was just brilliant.

I knew all the media companies were going to get into the Internet, but they didn't understand it. Jaime and I had as much of a chance as anyone, just the two of us, at doing something good. We understood the cyberculture and we understood the other stuff. I knew that once companies like Time Warner got involved, they could totally out-gun us in terms of information and the money they could spend. What they couldn't have was the flavor. They could never reproduce that flavor.

Eventually Icon realized they would not become millionaires instantly through the content business, and they got a grip on what their business actually was, which was hosting and stuff like that. They went public in '98, which was really the only goal of any of these companies at the time. And they had to file their first quarter public whatever-you-call-it, and wondered how they were going to justify the money they were spending on these webzines. And they couldn't really justify it. They hadn't been running them as a business. We had an ad sales guy the first year who actually sold half a million dollars' worth of ads and we only cost seven hundred fifty thousand to run, but they didn't want to invest. They weren't thinking of it the way real magazine publishers do—you lose money for five years, and hopefully you're building a cash cow that lasts many years. And they pulled the plug. They tried to sell it, but they were asking such ridiculous terms, out of pride. I had thought that the business world was a lot more sane than I am, and that's why I couldn't be in it. But now I realize a lot of these people are completely irrational. They're doing things out of emotional reasons and out of pride rather than for financial reasons. And I think that was part of my little business education.

*Word was shut down, but restarted a few months later by Zapata, a fish oil company reinventing itself as an online content network. Marisa reassembled the staff and relaunched the site.*

**MARISA BOWE**    One of the most amazing things about the past five years to me is just seeing how crazy people are. They have no common sense. And I realized that the fact that I had common sense put me far ahead, in terms of being able to accomplish things, of most of these business people with all their money and everything. People were starting these content models that couldn't possibly work, the way they were set up. Or their expectations were wrong, and then they would fail and people would say, "Content can't work." I mean, put a monkey in a car—are you going to say cars don't work, too?

But Word was never run as a business. We had no ad sales people, we had no publisher. We had me. And I was not a businessperson. Neither company understood how to run a media business. I did, but I couldn't do everything. I was basically trying to do things I knew a real media company would do, like make other projects that would extend the brand. We did a book and we did a game, both of which worked really well, but there was no support to do any more with that. I think of it as a band. And the band did all of the projects it was supposed to do together.

*Word folded in the summer of 2000.*

**CLAY SHIRKY**    To me, the day Silicon Alley ended as a thing separate from business in New York was the day Word.com folded. You could date it to the day Pseudo folded, find other dates, but to me there were a handful of cultural institutions that absolutely mapped the history of the Alley—that were forced to masquerade as businesses, because that's what one was if one was running a website. Word and Pseudo didn't go out of business because their business models were wrong. They went out of business because there *were* no business models. And when somebody finally said, That .com at the end of your name, that means *commercial*—why don't you come sit down with me and let's look over the numbers—[snaps fingers] they just vanished. It wasn't like they had a revenue shortfall. There was a set of numbers over here that was money going out, there was another set of numbers over here, I don't even know if those numbers were handled by the same people in these organizations.

Word was plainly a cultural institution. It was like a literary magazine that someone runs for the pleasure of running a literary magazine. Through some miracle, Word made it for five years. It was like Mr. Magoo, who, because of his nearsightedness, walks along, and just before he walks off the edge of a cliff a moving beam picks him up. Word was constantly being buffeted by outside forces. Marisa did an amazing job of piloting the plane from one spot to another. It was more like the *Perils of Pauline* than Mr. Magoo, I guess, because you never knew what was going to happen next. And they survived for years, and did really astonishingly good work. It wasn't of uniform quality, but every now and again you'd go and something would just knock you out.

**DOUGLAS RUSHKOFF** The content companies that collapse are not infrastructure companies. They're companies that have attempted to turn a communications medium into a direct-marketing platform. A great majority of the content that's been developed out of Silicon Alley has been one form of catalog copy or another. It's all been directed toward getting people's fingers away from their keyboards and onto the mouse. Getting people to consume information passively rather than actively contribute to conversation. It's not the kind of content that suits interactive media, and that's why it's not proving profitable or interesting.

**CLAY SHIRKY** This is my theory about why Silicon Alley as we knew it is effectively over. Word.com was plainly, on its own terms, successful. There was never any chance of turning that into cash. But in a world where for a few years their success had to be financial success, there was no way of saying, "We have a successful literary magazine, which loses money"—something the offline world has gotten quite used to.

*Nerve.com, the site for "literate smut," may have been late to the party when then lovers Rufus Griscom and Genevieve Field launched it in 1997, but their racy subject matter and propensity for publicity stunts launched them into the spotlight.*

**RUFUS GRISCOM**   We were the last of the webzines to emerge in New York. We launched in June 1997. I had consulted with Steven Johnson, with Omar Wasow, with Stefanie Syman, with Marisa Bowe and Jaime Levy, and became very friendly with all those guys. In effect I'd had the idea for many years, and I basically felt that there wasn't an intelligent magazine about sex for a female audience in addition to a male audience. And that whereas *Playboy,* we're told, had been a somewhat revolutionary force in the '50s and '60s, it had become this very conservative, ossified, eroding brand. The space the Playboy brand had occupied had not been filled; there was a brand-and-content vacuum. And there were a number of specific things that made it possible for us to do this. To begin with, homophobia had subsided enough to make it possible to have a magazine with pictures of naked women *and* naked men. (Whereas when I was in high school, no *way* would we have looked at a magazine with naked men in it. We would have freaked out.) Also, the country has become significantly more educated; the number of people who attend some years of college had grown from twenty percent to sixty. So obviously our sensibility comes from a much more educated, thoughtful direction.

Nerve has worked, and a number of other content companies have worked, because we were participating in cultural change, which is separate from technological change. We were taking advantage of both a cultural change and a technological change; we saw a cultural opportunity.

**STEVEN JOHNSON**   Rufus, Nicholas Butterworth, and I all roomed together in college, and Amanda Griscom, our first employee at FEED, is Rufus's sister. So when Nerve arrived there was a lot of fun back and forth, collaborating on what Nerve was going to be.

**STEFANIE SYMAN**   And Omar and I went to dinner with Genevieve and Rufus, having read the business plan, to talk about Nerve but also to talk about running a business as partners, because Omar had run New York Online with Peta Hoyes and that partnership had dissolved. And of course Steven and I had been doing FEED.

**RUFUS GRISCOM**   Jack Murnighan, who's editor in chief of Nerve online and went to school with me at Brown, got a Ph.D. in medieval literature at Duke. Steven [Johnson] was in a Ph.D. program as well, and a lot of these people thought at one time that they were going to spend their lives in academia. I thought I was going to spend my life in academia when I was in college, but obviously instead I started this with Genevieve. I brought Jack in, and then Joey Cavella, who is our designer and was my sister's boyfriend. I was running this business with my girlfriend, my sister's boyfriend, and my best friend from college, so it could not have been more fun. It was really kind of Utopian in those early years.

Once we saw this opportunity, the first step was to talk about the idea in the presence of rich people, hoping that eventually one of them would write us a check. After many months of talking about the idea in the presence of rich people, and subsequently negotiating with the lawyers of rich people, we raised about a hundred thousand dollars. It was not a sweetheart sort of deal; the original investor got a very large stake in the company, and the amount we raised would today be quite negligible.

**STEFANIE SYMAN**   Rufus hadn't quite made the transformation to Hugh Hefner yet.

**STEVEN JOHNSON**   His "Sex Media Mogul" presence hadn't yet developed. I've known Rufus since we were nine, so his whole presence has changed so much since that time. I don't really remember what stage he was in during the launch. Very excited, very energized, I think they knew they had a good idea—that was clear from the beginning.

**CECILIA PAGKALINAWAN | CEO, BOUTIQUEY3K |** We were creating an e-commerce site pro bono for Nerve in exchange for shared revenues, and I felt really defeated when that deal came to an end. The way their financial people were looking at it, they wanted one hundred percent of the profits. I was putting up all the work, and I was friends with Rufus, and that's when I saw that money talked before relationships and loyalty.

**DAVID LIU**   Nerve is definitely a company that is mentioned in conjunction with us [TheKnot.com]. But if we are talking about companies that are successful I wouldn't throw Nerve in there, because they're going to have to deal with a very difficult issue soon, which is the return on the investment they got from their original investors. I don't believe that there's a public market for their material. Playboy.com had a hard time getting out [going public], and I don't think they did—they pulled their offering. If Playboy can't get out then Nerve is not getting out. There is just no way. I don't think that those investors were very interested in creating a profitable private business—they're not looking for that.

I remember talking to Rufus and Genevieve, and saying, "Be very careful before you take your financing, because once you do your world changes. You have done a deal with the devil; those people who have invested in you may think that you are the most wonderful person in the world, but they are looking for a return on their investment." I think they had a bit of arrogance, too, because they thought they would rise above that. At some point they started ratcheting up the stakes, because they started raising more money. They wanted to grow more, and I think that's what left them hanging.

**ANNA WHEATLEY**   I recently picked up a copy of the Nerve [print] magazine. I don't think it's what they set out to do, but I think they hit on the right formula, which is to use their website to build an audience and then flip the brand into a multimedia platform. And it can extend beyond print. It could go to cable, or they could do a segment for the Playboy Channel. I respect the fact that they were here early, they plugged along. I've never been to their office, done an in-depth interview or anything, but the soft-porn aspects didn't bother me at all— pornography is a great business. I think going to the magazine format was the only way to go. And if they are successful—that will be the model.

*By the end of the year 2000, Nerve had launched sites in several different languages, published three books and a CD, had a thriving personals section, and had launched a print magazine—the first issue of which featured nudes of Tanya Corrin, then girlfriend of Josh Harris, the founder of Pseudo, and a former Pseudo employee herself.*

**CLAY SHIRKY**  There was a moment in 1997 when we said, Wait a second—in a world of supply and demand, if supply is infinite, price falls to zero. That's why you can't charge for content. It was a total mystery. Other people had figured that out; I'm not saying I was the first. But no one had ever published it that I read, and everyone was still banging on about "content is king," and suddenly I woke up one day and went, Oh, *that's* why we can't charge for content. It's just macroeconomics. Once people had figured that out, you didn't need to ask me or anybody else why that was true—it was just absorbed by the culture.

**KEVIN RYAN** | CEO, DOUBLECLICK | This is exactly what we thought was going to happen. To me it's no different from if you opened twelve pizza parlors on this street. It's not the pizza that's the problem, you just have too many parlors. What would happen over the course of two years is that ten of them would go out of business. You'd end up with two, maybe one upscale, one downscale, and they'd be very successful. Meanwhile, all the guys going out of business would say, "You know, these people don't like pizza."

That's exactly what's happening on the Internet. There are just sooo many sites . . . which is healthy and good. But remember the normal math of start-up companies. For a while people forgot that eight out of every ten businesses don't make it. That's just a reality. And it's going to continue here. What you'll see, if you do the math for any sector, is that if you reduce it down to three or four players it works fine. Because the volume doesn't go away, but the costs do. So you're going to see a lot of consolidation.

**KEVIN O'CONNOR** | FOUNDER AND CHAIRMAN, DOUBLECLICK |
There are a lot more profitable sites out there than people even realize.
There are thousands of profitable sites. The thinking was sort of get-
rich-quick, and basic business sense wasn't necessarily applied.
Whereas with traditional media it might take seven years to launch a
magazine and make it profitable. It took *USA Today* something like fifteen years to become profitable. It takes a long time for content companies to become profitable. And most of these Internet companies are
four years old, so it's going to take another three years for them to
become profitable. Can advertising be the only model? Maybe not, but
it's already a five-billion-dollar industry.

**KEVIN RYAN**    The best example, the precursor of this, was the story
of the search engines. Pull back the headlines in '97 about InfoSeek,
Lycos, Yahoo. Everyone said, "They're not gonna make it; it's
advertising-supported; it won't be good enough; they're not making
money." What do you see three years later? Because they were the first
ones out, they're almost all making money, Yahoo's making a ton of
money, Lycos is profitable, InfoSeek is part of Disney so it's a little bit
hard to tell, but Excite was not far off. There are tons of companies from
that generation now—in fact, the majority of the publicly traded companies that are four years old are making money.

I often read articles where people ask, "Why aren't these companies
making money?" In the same publication they say, Well, obviously, if
you launch a cable channel or magazine, it's seven years before you
break even. It's just understood that that's the average. Yet somehow
people didn't expect that to apply to the Internet companies. At the end
of the day, they're media businesses just like the others.

*Has content has been a failure?*

**ESTHER DYSON**    It's not just about having the right business
model, it's about implementing it well. I think advertising on the Net
is a fundamental misuse of the Net, as opposed to two-way communication and direct marketing. And you can do direct marketing
poorly—which is spam—or you can do it well, which is permission
marketing.

We need to find more things that are native to the Net, rather than simply take what works in other media and pop it over.

**JOHN YOUNG** | CHIEF CREATIVE OFFICER, TRIBAL DDB NORTH AMERICA | We have to figure out what's the equivalent of film edits online—how do you create surprise, how do you create sadness, compassion? Those are different levels we have to explore. Right now it's just about "graphics" and "content" and all these inhuman things. The person I want to kill is the person who thought of the word "repurposing." And that was the beginning of the Internet. In the beginning, everyone was selling it as "What's so great about the Internet is that you can repurpose content from offline." Content: inhuman. Repurpose: imitation. *Boring* is what it turned out to be. So we have to find originality, we have to be relevant, we have to have impact, we have to create talk value, find ways to touch society and individuals.

**ESTHER DYSON** I don't think there is a single "proper" content model. There are some that are improper, like giving everything away for free and leaving yourself no way to generate revenue, but I think there are more models that will work when done right. The model alone won't make the business. The content has to be compelling. Again, there's a lot more to this business than entertainment. There's doing useful stuff, such as recruiting and training people. What I don't like is this fascination with media, as opposed to useful information, such as journalism, and useful activities. As an investor I like things like debt collection, which happens to be very useful.

*What about your investments? Can content sites survive?*

**FRED WILSON** | COFOUNDER, FLATIRON PARTNERS | Absolutely, but you have to do three or four things really well to make it work. One is that you have to look at offline extensions of your business. It's a really weird thing, but people might not read the newspaper; they might read all the stories on the Internet; but for some reason it's easier to get a company to pay ten thousand dollars for this print ad than for a banner on the website. It's easier to monetize a piece of paper than it is to monetize a website.

Second thing you have to do is email, and push, and do that kind of stuff really well. You have to get the content out to people, and you have to get the email addresses from people, and you have to push the content through newsletters and things like that. And you also have to think about how to produce content efficiently with message boards and community-generated content and things like that so you can generate more page views per content creator than you might otherwise do. So there's a lot of things you have to do well; the companies that do all of them well will probably succeed, the companies that do some of them well will probably make it, and the companies that only do one of them well will have a hard time making it.

TheStreet.com is a good example. TheStreet.com does the journalism piece on the Web really well, but they haven't done a good email product; they haven't done any offline extensions, they haven't figured out how to do community very well, and as a result they're struggling. Someone like PowerfulMedia [Inside.com], they do online content well, and their magazine I think is going to be a big hit. So then they'll do two of them well. If they can get community right, then they'll be in great shape. You have to have the entire piece of the puzzle. But I believe the Internet as a medium is as well suited for publishing and journalism as any single thing the medium does.

**JERRY COLONNA** | COFOUNDER, FLATIRON PARTNERS | Can content companies be successful? I would say that from an investor's perspective the answer is absolutely, and from a company founder's perspective the answer is absolutely. The paths to success are in alignment, but the focuses are different.

From the investor's perspective, the way to be successful in investing in content is to pay the right press. It sounds simple, and it sounds simplistic, but one of the problems in the last two to three years is that the prices got ahead of themselves. And there was this notion of an Internet premium. The fact of the matter—and we've said this many times—is that a media company is a media company is a media company. And if the prevailing wisdom is that media companies get [valued at] three times [annual] revenues, then that's what you pay. That's the price of the business. As long as I, as a private investor, understand that and apply that metric, I can make a ton of money making investments.

Where the perspective got twisted was the notion that content sites could be and should be independent, publicly traded companies. Because, historically speaking, most media companies do not have a single property. They have a multiple property, multiple-audience strategy.

The last thing I would say is that demographic plays and understanding how demographics fit in is more important in this medium than in any previous medium. Because you don't have pass-along the way you have pass-along online. And you don't have the bullshit, frankly, notion of Nielsen ratings, where hundreds of millions' worth of advertising decisions are based on bullshit data entry. I need to know what a specific person's page views are, and I will pay advertising based on those page views. That's a phenomenon we've never had to experience before as journalists and editors and content providers.

**NICHOLAS BUTTERWORTH** We'd had a project called Indie-Net, which was a joint marketing group that included TotalNY, FEED, Agency, and SonicNet. A year after we founded it, most of the companies had been acquired and weren't really *indie* anymore. The culture had started to change. It was starting to be acknowledged that this was a business, it was a content business but it was still a business.

**JASON CHERVOKAS** There was an attempt to pull this all together at the IndieNet event, which was a disaster, but I think it captured the spirit of those days. It was a very small clique of people who were working on creative ideas, and the platform was shifting so incredibly quickly that no one really knew what to do, so everyone was sort of watching everyone else and trading ideas. It did not get really competitive until the money started to arrive.

**STEVEN JOHNSON** Some people wanted to make IndieNet a little consortium of independent content people. But I think the thing we most wanted out of it was to create an ad sales group network. Everybody else had their ad sales team, 'cause they were really less indie than we were; at that point we were really just three people. We just wanted somebody selling ads for us, and we didn't want to have to pay for them. But we also thought we could maybe do events together. So we put together this thing, and we ended up having these bizarre, ineffectual meetings. It was very ill-planned.

We never got the ad sales thing together, but we got ourselves together enough to put on a subconference within Internet World at the Coliseum. We invited all these other indie sites like Suck.com to come and showcase, and it was the first time we had met them at all. We'd sent off this stuff to Suck saying, "We're doing this big party at the Coliseum in New York," and it just seemed like it was going to be a big deal, so they brought over all this stuff, they had a refrigerator, and Suck magnets, and Suck-branded Coca-Cola, and they had a whole display of incredible stuff.

FEED literally had—we brought that one computer we had; it was still our only computer. We brought it up in a cab, and had a card table with our one office computer, and that was it—we were just showing FEED on it. It was over the weekend, and Stefanie and I both had a wedding we had to go to, so we had this intern we'd had for about three months man the booth, and the Suck guys were like, "What the hell?" They were used to these West Coast conferences, and there we were in this godforsaken little subwing of the Coliseum with our card table. So that was kind of the death of IndieNet, I think.

Everybody just thought, Okay, that was a total disaster.

# COVERING THE ALLEY

*At first, the growth of New York City's fledgling tech industry was largely ignored in the national media, passed up in favor of get-rich-quick tales from Silicon Valley.*

*In response, homespun media outlets like* Silicon Alley Reporter, @NY, *and* Alley Cat News *provided analysis and helped celebritize the entrepreneurs—most explicitly with the "Silicon Alley 100," Jason McCabe Calacanis's annual list of the most influential players in the game.*

*Alley entrepreneurs began to find a willing audience in the press, eager to catch tales of all-nighters at the office, raging dot-com parties, and the spending habits of the newly rich.*

*The media played a crucial role in the development of Silicon Alley, feeding the growth of Internet mania with magazine, newspaper, and television features on the irresistible young and successful entrepreneurs.*

**CLAY SHIRKY**   In '93 the Internet was starting to get hot. If someone mentioned the word "modem" in the newspaper, we would all rush to Panix and be like, "Hey, did you see? The *Daily News* used the word *modem* on page B17!" It's hard to believe that was only seven years ago. But anyone who was peeking under the hood into this world from the popular press was really extraordinary.

**SETH GOLDSTEIN**  It's based in New York; it had a technology angle to it; it was young, nineteen-year-old entrepreneurs: it had all the elements of a good story.

**MARK STAHLMAN**  Some of the established media companies have recognized that new media isn't necessarily their friend. As much as there are many individuals involved with new media in New York who don't really have a larger view of New York City life, you still can't really accuse people who run the media in New York of being naïve.

So the editorial boards of the *New York Times* or *New York* magazine or the *Village Voice,* to name a few that have built little empires here in the Empire City in the Empire State, are very familiar with how this stuff works. And they have been, I would say, uniformly hostile. For good reason: Magazines and newspaper and radio and television are all finished, in a certain sense, and they know it.

Television is the dominant medium of the last fifty years, so if you're a newspaper or a magazine in that context, you already know your number's up. Newspapers are a nineteenth-century medium.

I don't think the *New York Times* has ever done a serious story about new media in New York. Since the *Times* believes itself to be the newspaper of record, anything they put in the newspaper is on the record. The way you know the *Times* isn't on your side is when they don't talk about you. And if they're really unhappy with what's going on, they will simply just not put it in the newspaper.

**JASON CHERVOKAS**  As a journalist in America you're taught to be objective. Which means you're supposed to get a "he said, she said" quote and run that. And I think what happened with the birth of the Internet was that journalists typically don't understand math and science, and they didn't really understand business and technology. So someone would tell them that some new company was the greatest thing since sliced bread; then the publicist would feed them to an analyst who was on the payroll—getting a consultant fee—who would say,

Yeah, this really is the thing. And suddenly it would get overplayed on A1 of the *New York Times,* and we were off and running.

*In September 1995, Jason Chervokas and Tom Watson, two local reporters founded @NY, an email-based newsletter about the industry. They eventually sold the company to Alan Meckler's Internet.com in the Spring of 1999.*

**JASON CHERVOKAS**   My partner Tom Watson and I were working at the *Riverdale Press,* which is a community weekly in the Bronx. We had been independently interested in, and using, the Internet in a variety of ways. Covering politics in the Bronx has always meant covering industrial policy issues. And industrial policy development in New York City has always meant job retention—big companies saying "I'm leaving town," and the city saying, "We'll cut you a tax break to keep you here." We just kind of fell in love with the story of what was happening with a handful of Internet start-ups that sprung up for a bunch of accidental reasons. And no one was covering it; no one thought it was a story. If it was a story at all, it was kind of a joke for West Coast journalists. But it was really fascinating to us.

We also fell in love with the notion of email publishing. When we launched, the graphical Web browser was still a bit of a novelty. But it was a novelty that was very much the hype of the day. The Louvre website was actually the big, popular website of the time. It was international, it was art, and you could go to it. And we were coming from a very old-fashioned community newspaper business, and thinking that email publishing was a way to do printlike publishing—shifting the cost of print onto the end user, eliminating the cost of distribution. Wasn't this going to be the way that all printlike media was going to be distributed? And furthermore, as publishers of a niche publication in a format like this, where we knew something about who we were sending this to—didn't we know more about our users than anyone who was Web-obsessed?

I really wanted to do a little more business planning before we launched @NY, but Tom decided, Hell, let's just launch this. And we wound up spamming a couple of Usenet groups in August of '95, and suddenly we were on the hook for delivering an issue September 1, which we really hadn't plotted out, but we knew something was happening. We produced it as a biweekly newsletter on Friday nights after work for a year. That was the origin.

*In 1996, Jason McCabe Calacanis, a columnist covering the digital scene for* Paper *magazine, published a photocopied newsletter called* Silicon Alley Reporter, *which would evolve to become a full-fledged mini media empire, with a sister publication covering the tech scene in Southern California, several annual conferences, and a suite of email newsletters.*

**JASON MCCABE CALACANIS** There were probably about five or six journalists who were covering the beat: myself, Jason Chervokas, and Tom Watson, one or two other journalists who worked for the New York [*Daily News*], like George Mannes, who eventually wound up going to TheStreet.com. Saul Hansel from the *New York Times* started to get it; he was covering bigger companies. There were four or five of us out there, and @NY had the guts to start this weekly newsletter. And then I started the print magazine, and people thought it was very odd to create the print magazine, but I based it on something very similar, which is Esther Dyson's *Release 1.0*, which had become a very big deal.

I started doing the math about how much that would cost. I said it's going to cost, like, two or three thousand dollars to print this thing up. Then I went to print it up, and I realized I had to photocopy both sides of the page, so it doubled the cost. That really was a lot of money. So I put about ten thousand dollars on my credit card to print these issues up.

I used a photocopy store up on Forty-third Street. I remember going there at night with the Zip disk and talking to the guy and I convinced him to fold it for me and not charge me. I made friends with the night clerk. I wouldn't go during the day, because if I went there during the day the manager there would have charged me. I just hung out there as it came off the press.

[When it came to soliciting ads,] I took photocopies of what would become the *Silicon Alley Reporter* and I asked one of the designers I was collaborating with to make these third-of-a-page ads that go down the side of the magazine. I saw it in *Paper* magazine and I thought it was a clever idea. I didn't think anyone would buy a full-page ad. None of the

Silicon Alley companies had done any [print] advertising before. They didn't have some ad sitting around waiting to run in this magazine.

With my mock layout in hand, I thought I'd get everybody to commit to four issues in advance at two hundred fifty dollars and pay me a thousand, and the four thousand would pay for the first print run, maybe. So I went to Connors Communications, Jeff from Razorfish, Adeo Ressi from methodfive, and Brian Cooper from Ernst & Young. These were people I knew from my column, or whom I'd mentioned.

I was at the Connors Communications Christmas party with that piece of paper, and I took it out and showed it to Connie Connors and a couple other people, and other people started walking around the table. And I sold four or five more ads right there on the spot. All of a sudden I had nine ads in the first issue—double the revenue—and I said, Wow, there's some market for this.

Then I got the check from Jeff Dachis, and it was for two hundred and fifty dollars. So I called him up. I said, "Jeff, remember? I said to pay me a thousand." He said, "Well, I want to see how it goes." I said, "Jeff, but if you give me the seven fifty I can print up another seven hundred fifty issues, and more people will see the ad." He said, "Oh, that makes sense," and he sent it the next day.

I always had stacks of issues because I couldn't afford to deliver to anybody. I decided to deliver the first subscriptions by hand. So I went to Dan Pelson's office and knocked on the door.

I said, "Is Dan Pelson there?"

"Yeah, I'm Dan Pelson."

"Oh, here's your *Silicon Alley Reporter*—you subscribed."

He said, "You didn't send it by mail?"

I said, "No, I felt like giving it to you by hand."

That became sort of my beat. I would take four or five hundred copies of the magazine, put them on a luggage cart, and walk around Silicon Alley dropping them off at people's lobbies: SiteSpecific, Razorfish, Pseudo, whoever it happened to be. I'd say, "Do you mind if I put this in the lobby?" And they said, "No, great." Everybody would come to the lobby and grab them.

**GORDON GOULD** | FORMER PRESIDENT SILICON ALLEY REPORTER |
Physically, *Silicon Alley Reporter* was a black-and-white Xerox. It was a

rag, basically, but it was a rag we all loved, and we all read it. It's evolved from just being a raw 'zine-rag into something you can read to learn about who's who in the industry and what's going on, and what some of the interesting and emerging trends are. Jason is very prescient in his understanding of the Internet, and I think people appreciate the fact that he's highly opinionated. *SAR* is not really an objective information source; it's not trying to be an AP or anything like that. But I think people like the opinions it puts forth. They might not agree with them, but they appreciate the opinions.

**JASON MCCABE CALACANIS**   Jason Chervokas and Tom Watson [of @NY] were true journalists; they went to school for journalism. I was just a hack.

**CLAY SHIRKY**   I remember when *Silicon Alley Reporter* was a little gossip rag. And you'd go pick it up, thinking, What is this thing? It's so crazy. It was really like a Liz Smith column: You'd read along and there'd be Chan Suh's name in boldface, and you'd wonder, "Who does this Calacanis guy think he is?" He's such an amazing story, because he's totally a home-grown guy who saw a chance and said, "Nothing's standing in my way." It's astonishing what that's done. And largely what it did was prove out the viability of that model [online, in print, in person], to the point where the *Industry Standard* could be founded.

Interestingly, at the same time as *SAR* was coming out, *Wired* was coming out, with its kind of Utopian thing. And it's fascinating to me that in the five years of their covalent history, *SAR* is in some ways the more viable publication, because it's covering a real business community and has spread out. *Digital Coast, Pervasive Weekly, iHealthcare Weekly*—Jason's got a little empire there. And the Louis Rossetto [of *Wired*] approach—We're going to take this revolution to the masses— they look like the *Omni* of our generation now. You open it up, and there's some big, big project; they're digging a giant hole somewhere, they're laying cable across the sea, and you think, I thought I picked up *Wired*! What's going on? Meanwhile Jason's writing, *So and so moved from this business to that business and now they're heading up this*. Oh, that's interesting.

**NICHOLAS BUTTERWORTH**   Microsoft had this new Microsoft Network service, and they were going to have the best content, and that was going to be their differentiator. And they had this guy Bob Bejan who was a Hollywood guy, and he was going to have a studio model where they were going to have pilots and then put them in development, and they'd commit and give them distribution and ancillary spinoffs to TV and other media yadda yadda yadda. When I hear people saying that the reason their website will be successful is because of their spin-offs in other media, I've learned to get really suspicious, because usually they should be in that other media to begin with. It's okay if it's the secondary thing, but if that's the original bet then you inevitably have a problem.

So Lara Stein from MSN came to New York, and she did what people in development in Hollywood do, which is go around, meet with everyone who's creative, and lead them on like crazy. And that's what she did, because it was her job and she was good at it. And she met with everyone in New York. Everyone raced to show their best ideas, and I think they ended up green-lighting one project—Tim Nye's, which they paid for development of and never launched. And I think that's it. Out of maybe a hundred, two hundred pitches that they took from everyone in town, maybe they paid for one and launched zero. So Jason basically wrote a story saying, This is what's going on: Microsoft is pretty unpopular because they're scamming. And it didn't endear him to Microsoft, but it did get him a lot of attention.

**JASON MCCABE CALACANIS**   In the first issue, besides saying NYNMA was out of touch, I also said that Microsoft was upsetting people in the Alley by trying to own all the digital rights to their work. With those two stories coming out in the first issue—two controversial stories—I mean, everybody wanted to read it. And then probably six weeks later the *New York Times* ran their own Microsoft story featuring Lara Stein. The fact that we did a story on this one particular person, who nobody knew before that, and then a few weeks later the *Times* did a story covering exactly the same thing, suggested to most people in the Alley that they got it from us. That's when we came on the radar—when people realized we got the stories first.

**GORDON GOULD** Jason's conferences actually started because he had done some informal events before I had joined the magazine, but I'd been bugging him. I said, "Jason, you know the magazine's a good thing, but you can make a bucket of cash doing events." He said, "Well, why don't you come and do the events, then?" So I started the events division at *Silicon Alley Reporter*. But it was something that has been—and will I'm sure continue to be—a cash cow for the company.

It's very prestigious to be in print, because there's a finite amount of space and it's something you can hold and give people. So there's a lot of perceived value in being written up, especially being on the cover of a print magazine. You really use that to push your flagship identity. The online component of it is more about immediacy, about what's happening right now. And then the in-person component was the place where people could really make it all come together for themselves. They were aware of what was going on in the industry, they knew who the players were because of their pictures in the magazines; there was a certain star quality that was bestowed upon certain entrepreneurs just because they had been in print a lot. And now you could come to the conferences and meet these people.

The dirty little secret about conferences is that, yeah, the content's important, but it's really very marginal. You need a good enough content billing to get people to come, but what you really need is the right audience. So if you give them sushi, you give them a great place to hang out, and you give them a forum to interact and discuss business with each other, you will make a lot of money. Times are a little different today because the market's so crappy that people are a little more utilitarian, but there's still a basic need and desire to rub shoulders with industry leaders and to go see and be seen.

**JASON MCCABE CALACANIS** I was part of the culture. I was out every night. That's what differentiated me from Jason and Tom, and so that became the battle. It was @NY versus the *Silicon Alley Reporter*, the weekly email versus print. The journalists versus the, you know, hack. Everybody needs a foil. Everybody needs a competitor. Coke needs Pepsi. And we both rode it for all it was worth.

I realized that, and then *New York* magazine came to me and started talking about it. I said, This is going to be good, if I can pick a fight with these guys and have it get a lot of press, because nobody knows who we are. I said, "Those guys will never really amount to much because they are not part of the Alley. They have to go back to the suburbs at night to take care of the wives and their kids, and I stay out all night. I'm part of the culture. That's why they're never going to matter."

It was a very immature thing to say, but it was true. And it was a great quote. So I got the pull quote in the story and they didn't; I got the final word and they didn't. I started realizing how you play the press game, and how you get the press to write stuff about you. And the public persona of Jason McCabe Calacanis started to grow. It's probably not that close to who I actually am, but you don't really have that much control over what the press does with you. So the press made me into, you know, the insider. Fine, okay. I'm the insider if that's what you want me to be. It sells magazines.

That was the thing about Jason and Tom: They were clearly better journalists than I was—*clearly*—but they had no idea how to market, and they didn't live their brand, and I did. And I did print.

About a year and half or two years into it, people started saying *Silicon Alley Reporter* is cooler and hipper and I love the parties, but @NY is better journalism. And I just got tired of hearing it. So I spent the money and hired really good journalists. I just went on a tirade looking for anybody with degrees from Columbia Journalism School. I had to hire people who knew more than I did about journalism and learn about this stuff.

I've never really talked about this, but I said, Okay, how do I one-up @NY? Because people would play us off each other. What would happen is, I would do a story on somebody, and I would need three or four weeks to get it out in print. A week or two after I would talk to them, they would go talk to Jason and Tom. Jason and Tom would get the story out the week before I came out in print. So competition happened.

So I sat there trying to figure out how I could do something to trump @NY, because they were just a thorn in the side. And I said, I'll do a daily email. And I'll do it in HTML, because the new email supports HTML. And I'll put pictures in it. And then I'll scoop them.

I always make the analogy that the Knicks always got better when Michael Jordan came to the Garden. The Knicks would be incredible:

the defense, their offense, everything rose in the game. I don't know who was Jordan and who was Patrick Ewing in the equation, but we made each other better for the competition.

**JASON CHERVOKAS** We are the only Internet start-up of the three—*Silicon Alley Reporter, Alley Cat News*, and *@NY*—to actually have sold a business. The others are still waiting for their liquidity events. In our time running *@NY*, we cut content licensing and business development deals with AOL's Digital City group, the *Industry Standard*, and the *New York Post*, among others. So we weren't pikers when it came to building a media business. And in the final analysis we sold a year before the market crashed. In fact, our timing was impeccable.

**GORDON GOULD** Jason has a real talent for understanding what people's base motives are. And people like to be starfuckers. Look how popular *People* magazine is. *Silicon Alley Reporter* in the earlier days was like a *People* magazine for the industry. And you could see *yourself*, and people you knew, in this magazine. That was very appealing. It's much easier for the media to get excited about a person than about a concept.

**ANDREW RASIEJ** The *Silicon Alley Reporter* 100 list was a milestone because it started to celebritize individuals in the marketplace.

**MARC SINGER** It was cool to make the *Silicon Alley* 100. I think we were in it for like three years in a row, and the first year nobody cared. *We* cared; we were like, It's so great that those guys like what we're doing and see that we're trying to develop something really different from what's out there. As the Alley grew up, this past year, suddenly you go to the photo shoot and Sam Donaldson's there . . . There's intrigue. People are calling me from high school because they see a picture. It's weird. Early on it was funny because it was small. Now it's taken much more seriously.

**CLAY SHIRKY** The signal innovation was the *Silicon Alley* 100: That was the moment. You hated yourself for looking at the list, you hated yourself for thinking you should be on the list, but you had to see—were you on the list? And if that person's on the list, why aren't I?

Nobody liked it, except maybe the people who were listed first, but everybody looked at it. You would see the list and think, How many people here do I know? It was a kind of measure of connectivity. A lot of people hated Jason for making it, it seemed divisive and so forth. But as so often, he understood what would happen. The industry was too big for there not to be some arbiter, and since nobody was standing in his way, he was going to be that arbiter.

**GORDON GOULD**    Lists are nothing new, but the list was *wielded* at the *Silicon Alley Reporter*. People were very desperate to get on that list. I've been asked by the CEOs of public companies—begged—to make sure their position on a list is higher than a competitor's. I mean, people took it very seriously, and I think they still do.

People like to say *Silicon Alley Reporter* is incredibly biased, that it's just about friends of Jason, but I've always taken issue with that. Clearly it started that way, but we were very diligent about growing the circle of companies that got coverage, and Jason was very open to that. Jason is an editor, and he's a flamboyant editor. And he's an opinionated editor, so people will take pot shots at him. But the list sells a lot of ads, and seeing as I'm still on the board of the company and have a piece of it, I'm delighted that it's as effective as it is.

*The Silicon Alley 100 brought together the leaders of the industry, but it also never failed to hurt somebody's feelings. Josh Harris, founder of Pseudo.com, was particularly upset when the 2000 list featured a photograph of Larry Lux, his recently departed CEO. For weeks thereafter in the media—and once in public at a NYNMA panel discussion—Harris missed no opportunity to needle SAR's editor, Jason Calacanis.*

**JOSH HARRIS**    First of all, I love Jason; I respect him greatly. But if he can't take a joke, screw him. The real problem, I think, is that when I was around him, he just kept striking out with the women. So I think there was this subconscious transference, where he just screwed me in the Silicon Alley 100 instead.

**JASON MCCABE CALACANIS** Pseudo had management changes right as the issue was going to bed, but we couldn't stop the presses to switch the photo of the CEO. Josh is always looking for something to rub me on, and I think he had a little bit of egg on his face about Larry Lux. He fought to bring him in there, and not too long after he was gone. So I think he didn't really need a reminder of it or for everyone to see it and be reminded of the situtation.

**SETH GOLDSTEIN**   Jason was the Steve Rubell of Silicon Alley. It was like a really hot club for a couple of years, and he was the guy at the rope. It's increasingly irrelevant now. There's not enough action, volatility, to make anyone any money. It's not interesting anymore. The list is propaganda. More and more desperate. This year is going to be like, who lost the least?

**CRAIG KANARICK**   Jason was a pain in the ass, straight up. I love the guy, too; I think he's great, and he deserves an enormous amount of credit for the success of the whole industry. He was absolutely the most faithful cheerleader and believer who wasn't actually doing the work. We of course were saying this was going to be big, but we were *doing* it. He was saying it as an outsider. *Outsider* is maybe the wrong word, but as someone who wasn't just trying to sell his services. And that was fantastic.

He was relentless. I remember him walking around himself passing out the papers. But he covered my personal romantic life in his gossip column. I remember at least once when he saw me outside after a party with my girlfriend [Rebecca Odes of gURL.com] and tried to capture a picture of the two of us because we were both in the industry. The two of us were turning around going, Fuck you—go away.

On one hand he was great; on the other hand he was a pain in the ass. But that was part of his charm. We were *all* a pain in the ass, and we were all brash. It was all about being brash and being different and being new and being independent and being punk rock and being loud and being, *Fuck you, Mr. Big Company, you don't know what you're doing.* Thank God he built all this hype around the industry, because it got a lot of attention for it and helped make it real.

**DAVID LIU**   Jason Calacanis is the perfect example of a parasite who has made a business off of tracking this industry. And you know what?

That was smart, that was very smart. He's been able to leverage himself into being this little curmudgeon—he is the bellwether for the industry. He's turned his magazine into a significant publication. He is a good example of a smart New York–edged entrepreneur who has capitalized on Silicon Alley.

**DOUGLAS RUSHKOFF**   I think Jason loved it that multimillion-aires were competing to get on his list of top executives. This kid who came out of Brooklyn with a stapled, Xeroxed newsletter, he looked at this and realized that these companies are so based on hype that he could create a truly virtual pyramid, a list that meant nothing but that. He went meta on them. And these people who would not even let him walk in the door would be handing him cigars. I honestly believe he sees it as a joke. I admire anyone who's having genuine fun, and anyone who has enough perspective on this scene to realize that it is pure hype.

**CELLA IRVINE**   I think Jason Calacanis, Tom Watson, and Jason Chervokas have done an amazing service to Silicon Alley and the industry in New York. They have been fabulous in a whole bunch of ways: in promoting the industry, in creating that little bit of pressure that exists when you know you have to be interviewed by the media and you have to sharpen up your act.

*Anna Wheatley and Janet Stites, two writers working at the science magazine* Omni, *launched* Alley Cat News *at around the same time, with a focus on the financial aspects of the Alley.*

**ANNA WHEATLEY**   In the Fall of '96, Janet and I started *Alley Cat News*. It was a sixteen-page newsletter we produced ourselves, and our premise was: Wouldn't investors just love to know who these companies are? None of us were coming from a business journalism background. It didn't even occur to us to think about IPOs or mergers and acquisitions at that time. We were just thinking that these companies were young; they needed funding. We didn't know about the obstacles, but of course you're educated very quickly. And we knew on some intuitive level that even if *Alley Cat* didn't make it, people would pay us five times what we were making as freelancers because we were at ground zero of this phenomena. We weren't focused on the new CEOs, we were focused on the investors and the professional services—because in New York your lawyer and your accountant are the first grown-ups, the first mature advisors who come into a company with any experience.

*Of course, the presence of competitive news outlets sparked some resentment . . .*

**ANNA WHEATLEY**   I am very happy to go on the record to say I don't know any journalist in this town with whom I don't have a good relationship . . . with the exception of Jason Calacanis. He has been, from the beginning, antagonistic. With Jason I can't be the least bit objective; he has attacked our company; he's told people that we are going out of business; he has told both me and Janet that ourselves, in public. New York is big enough for competitors, and having two of us around has made it easier for both of us. When he first started his newsletter, he ran a picture of me in the paparazzi section over the caption "Anna Weenie." And I said to Janet, "That's an accident; he doesn't have a fact checker or a copy editor—he has lots of spelling errors." I mean, we did, too, but he had even more. Janet was like, "That's totally *not* an accident."

**JASON McCABE CALACANIS**   But the *Alley Cat* girls never really got taken seriously. I had a pretty standard joke: Anytime someone would mention *Alley Cat* to me, I would say, "What is that?" They started as a sixteen-hundred-dollar-a-year newsletter, and then it went to eight hundred dollars, and then it went to free, and they couldn't make it work, and their conferences couldn't work.

People knew I was chummy with Josh Harris, or with Razorfish, et cetera. So people sometimes critique me for being friends with those people and writing about them. But nobody would argue that Razorfish and Pseudo didn't deserve to be on the cover.

If you survive, you will be successful. *Alley Cat* will be gone this time next year, we won't. And then what happens? *Silicon Alley Reporter* is the only one left. Everybody wants to advertise, people still want to read about [the Alley], and I won't have any competitors. The race goes not to the swiftest but to the one who keeps on running. It's a marathon, not a sprint. Anybody can sprint, right? You can be totally overweight, totally out of shape, smoke two packs of cigarettes a day, and sprint a hundred yards. It has nothing to do with anything. It's who can run from Staten Island to the Bronx and then into Tavern on the Green. That's who wins.

[Wheatley and Stites] spent a good amount of time badmouthing us, saying we were too lifestyle-based. Four years later, they put a model on their cover in a skimpy dress. For them to critique me about seriousness when I put more [powerful] women on the cover than them is ridiculous. I've always said that lifestyle is a big part of Silicon Alley. To deny that there's a culture here is absurd. And they were very against the Alley culture. They were constantly down on entrepreneurs for having parties or being social. At one point they were upset about the Word.com party because they had dancers there. What's the big deal? There could be dancers at a *Talk* magazine party and nobody would say anything.

And it was funny, too, because the *Alley Cat* women always hated me, hated me, hated me. They're obsessed with me. They can't stand me. And the reason was very clear: I was better than them, and people knew our magazine. Also, they wrote a story about how Silicon Alley had to stop having parties, and they specifically put it on me. They wrote a big story about how on a human resources level all of these Silicon Alley companies were running a big risk by having all these parties, having strippers at the Word.com party, or dancers—it was go-go dancers they had, not strippers. It was a direct attack on me. They started badmouthing me to advertisers, and it worked with one or two.

**ANNA WHEATLEY** I wouldn't call it badmouthing Jason. When advertisers would ask us about *SAR,* which is a normal thing for them to do, our standard response was, "We're really not the best people to ask, because we don't have a professional relationship with them." I can't say, "Don't advertise in *SAR*"; that's never been our approach. Our approach to business was to be honest with our advertisers. *We* can't answer the question objectively, but we can tell you they are two very separate projects, and we are the quality product; we were trying to sell our product based on quality.

**JASON MCCABE CALACANIS**   And that's when I stopped doing the party pictures. I took them out of the magazine. I just said, you know, if that's what people expect me to be, I'm going to be what they don't expect. I'm going to be the best, most professional, well-designed magazine.

**CLAY SHIRKY**   If I were twenty-three, and it was 1997, *SAR* would probably be how I would learn about the industry. I can look at it and laugh and say that's Jason's take and here's my take, but a lot of people had no take on Silicon Alley, and Jason was the sole source for that information. There's also @NY, *Alley Cat News*; I've never really taken those guys seriously.

*By the time Nerve.com launched in June 1997, the national press had begun to really pick up on stories about young entrepreneurs, and the Alley presented many great opportunities.*

**RUFUS GRISCOM**   We happened to be able to launch on the same day that the Communications Decency Act was dealt a fatal blow, which was great timing. As a result, the day that we launched I was on CNNfn; a week later we were in *Newsweek*; a month after that there was a full page on us in *Time,* and off we went. We almost immediately had several hundred thousand readers. And in retrospect, arguably the single largest benefit of being online has been the press hype. What happened during this period was that tens of thousands of column mentions in newspapers, and just as many minutes of television time, had to be filled on a weekly basis with Internet-related stories. That's been the case for a few years now—the media has this massive, unquenched appetite for Internet stories. And if you could position yourself as an Internet story worth writing about, there was a huge upside.

In retrospect, you can look at Nerve as a magazine and as a media company rather than as an Internet company. If you look at the business model, they're not that different. The biggest cost is customer acquisition: if you can get massive free advertising, you've eliminated the largest expense. I'd say that the benefit we got from the hype was worth even more than the money we saved on printing or distribution, arguably. Of course, if you're talking to the press about it, you don't want to tell the press—though I do anyway—that they're part of your marketing strategy, that they're in your business plan, and that their totally kind of irrational obsession with the Internet is part of what has made it possible and has accelerated it. It's somewhat of a self-fulfilling prophecy.

**JASON MCCABE CALACANIS** If you knew what the Internet was in '94 and '95 and you were stupid or smart enough to start something—probably equal parts—you had a chance at either raising money or getting something off the ground and getting a lot of press very easily. I became "the Internet kid" to the New York press almost overnight. And if anybody wanted to talk about the Internet they called me, and to this day it is still part of what I do. It was hysterical, because if somebody from Germany came through New York, and they were doing a story about the Internet in New York, everybody would say "Go talk to Jason," and I would say "Okay." Or if somebody was doing a story about cafés and cybercafés—"Okay, go talk to Jason." And I'd say something about cybercafés.

**THERESA DUNCAN** One of the best things about being here in New York is being so close to the media. It's easy to talk to people about what you're doing, and it's easy to promote yourself that way. And I think it helps a lot in terms of selling things. I don't have a PR agency, so I have to do it myself, and if I were in Detroit or somewhere else it would be harder to catch people's ear.

**JASON MCCABE CALACANIS** I was a kid from Brooklyn, you know, and my mother had a subscription to *The New Yorker* and *New York* magazine and the *New York Times,* and in Brooklyn the only way you wound up in those things was if they took your picture when you were walking by on the street. The only way you wound up on TV was if at a baseball game you caught a baseball, or if somebody was doing a report from outside a police station and you jumped up in the back. And then all of a sudden, the *New York Times* put me on the cover of the business section. I said a couple of crazy things in the opening paragraph about how I started the magazine in New York because I didn't want to start one in Silicon Valley, because people like Larry Ellison and Bill Gates didn't trust me, and they didn't take good photos, so I couldn't do a good cover photo. I said that as a joke, and Amy Harmon printed it as a joke, but that piece started a sort of media barrage.

**NICHOLAS BUTTERWORTH**    Paradoxically, the Internet itself and the dot-coms became a much bigger part of the culture at large. The media take on Silicon Alley in '95 was, "Look at these freaks doing crazy stuff," but by '96 it had started to become, "Hey, this is the new way of writing." It was unbelievable how traditional media fell over themselves to praise and cover Internet companies. Every day we all got streams of visitors from traditional media trying to write positive things about Web companies because their readers were demanding it. It sold magazines and made them look hip.

There was journalist envy, where people were covering you, but they kind of wanted to be doing what you were doing. And the Internet became a place for former journalists to flourish. And with that came a move away from the original art and creative culture to something more business-driven.

**TODD KRIZELMAN | COFOUNDER, THEGLOBE.COM |** For us, the media coverage was something that was not very accidental. It was very much planned. And for a very, very long time it was a major source of our traffic. The reason people heard about TheGlobe was because of the press we had.

**ANNA WHEATLEY**    Janet and I were featured in the *Wall Street Journal* interactive edition right as we were launching. I was dying. I thought, Oh my God, we're going to be so successful. I can understand the temptation to believe that, because we are media, so we have a context for understanding it. But if you are not media and everybody is calling you up and they want to feature you on the cover of their magazine, you don't have any other way to interpret that, unless there is someone else saying, "Calm down, this isn't going to last."

**DAVID LIU**    When you're in school you think the world is full of people who are smarter and more experienced, and you trust what you read in the *New York Times*, what you see on CNN. Then you get into the real world, and you realize that this is all being created by PR people, and the people who are writing these articles are people who are younger than me and with no experience in this.

**RUFUS GRISCOM**  When we launched our site, my big concern was that I wanted to feel like I'd accomplished something in my twenties. Then a *Newsweek* reporter started writing an article about us, and I was waiting to find out which issue it came out in. It came out a week before I turned thirty, and it read, "Rufus Griscom, twenty-nine . . ." and I was like, "Yes! Yes! I barely made it!"

**JASON MCCABE CALACANIS**  One of the things about me that was endearing to the public, to the New York press, was that people were watching me mature. In the beginning I was twenty-four, and I was this impish kid running around creating a lot of noise and banging a drum. And then all of a sudden I was thirty.

**ANNA WHEATLEY**  I just saw that Courtney Pulitzer was featured in *Mademoiselle* as one of their "divas." Do you know that journalist called me looking for divas and they had to meet certain criteria—and they had to be under thirty? I can't tell you how many calls like that I get. The editors have a lot of pressure to do the youth culture angle. That's the image that's presented.

**THERESA DUNCAN**  The media was never very hard to court, let's put it that way. Being a young woman, people tried to focus on the glamour aspects, and that became uncomfortable after a while. Because I really started to get asked "What shoes are in?" and things like that. Celebrity was kind of easily coined, and all of a sudden a magazine would call me and ask me—along with a fashion model and a television or MTV personality—what I was buying that spring. And I really do like fashion, and I'm interested in theorizing about fashion, and I'm interested in it as a semiotic system and all this stuff, and I'm also just interested in it for form because it's good-looking and fun. But after a while that became uncomfortable. The media wanted to focus more on the parties and what I wore more than the work that I did, which was very strange.

**RUFUS GRISCOM**    I think that the example of people who appear to be making great fortunes and simultaneously living their dreams and contributing culturally was kind of devastating to a lot of people. You see a real appetite in the media now to portray Internet entrepreneurs as opportunistic, greedy assholes. And I understand why—it makes perfect sense. On the one hand the public wants to believe in the incredible good fortune and these beautiful lives they live, in the same way that people live vicariously through movie stars. And at the same time there's a real desire to demonize these people.

*One example was a March 2000 article in* New York *magazine, which many felt presented the early entrepreneurs ("Early True Believers or ETBs") as vain, pompous, and overhyped.*

**RUFUS GRISCOM**    Everyone knew the *New York* story was coming; it was pretty crystal clear. At that point everyone was pretty jaded about the press. We'd all exchanged emails about it. I would say that some of my colleagues have played into the hands of journalists looking to fulfill this need that the public has for demonizing Internet entrepreneurs as opportunistic snakes, and I think it's too bad. And I think it was definitely reflected in that article. People sounded incredibly puerile, childish, and petty. I was misquoted in that article in an extraordinary way. I said, "It's [powerful] to feel [that] you [are] one of seventeen people who understand the world," and I had written that in an email about my experience of being into cultural theory—capital T— in college, and it was extracted from a paragraph in which I was talking about my collegiate experience to put it in a context actually suggesting I was saying that about being an Internet entrepreneur.

**FRED WILSON**    If the media builds you up, the media's going to bring you down. Every single time it happens that way.

**JERRY COLONNA**    I think there's another phenomenon, too, especially with Silicon Alley companies. So many of these companies were launched by or staffed by renegades and folks who left jobs at traditional media companies. A lot of the people who stayed at the traditional media companies felt foolish for a very long time. For a year or two, they said, "So and so is over at this Internet company, kicking butt, and some day she's going to be worth millions, and I'm sitting here on my hands, waiting for editor number five to die so I can move up a slot." And I think that when the wheels started to turn in the spring of 2000, a little bit of the resentment became glee, and it was kind of "Ha, ha, ha, we told you so." And there's an element of that we-told-you-so going on in the stories.

**FRED WILSON**    The other thing is that people write about the obvious. It's easy to write the story about the fiftieth company going bankrupt, or the fiftieth CEO that got fired. They keep wanting to write that story. You get tired of reading it after a while, but it's an easy story to write.

**JERRY COLONNA**    Well, who put Jeff Bezos [of Amazon.com] on the cover of *Time* magazine as the man of the year? It wasn't Amazon's PR agency lobbying for that—you know they had no influence over that. It was the same traditional media people who at the same time now are vilifying this man, and neither is correct. He's an entrepreneur trying to build a business.

It's wacky. It was out of proportion twelve months ago in terms of the media hype, media frenzy; and Silicon Alley companies, if they've suffered, have suffered from being in the spotlight too much. You had relatively small companies with relatively few employees and relatively few revenue dollars seeming to have extraordinary influence, because this became a fascinating story within the overall story of the New York City economic boom. Live by the sword, die by the sword— I think that's exactly what's going on.

**CRAIG KANARICK**    I'm still learning about the press, and it is still not clear to me whether the press leads or follows what other people are saying. Do they read the press releases, or interpret what is going on? There was this fantasy story of kids getting rich quickly, which appealed to some sense of fantasy that people had. And there are people who used to be rich (though they really didn't have anything) who are getting punished for it. This is compelling to a lot of people.

There is this amplification of these people who supposedly didn't do any work and didn't have any skills or experience and had a ton of money. When in fact that was inaccurate for a lot of people. A lot of people did work incredibly hard, did have skills and didn't really have that much money. It was just on paper. They made it out to seem like it was more than that.

They told the up story, and now they're telling the down story, but pretty soon it will be boring. We didn't have a national scandal at the time—we were in between the Monica Lewinsky affair and the national elections. The Internet became the big thing between those two things that the nation focused on.

**LISA NAPOLI**　In the early days, Razorfish was everywhere. That's because press breeds press. I would talk to little Web start-up companies and they would say, "Why is it that we always see Razorfish cited?" And I said, "Because Razorfish gets in one story, and then everybody writes a story about Web design, and they go to Razorfish because it's just a no-brainer. Not because Razorfish is any better than anybody else—just because they keep getting their name out there." Part of that is because they were early entrants into the field.

**ALAN MECKLER**　Well, the press's job is to report what's in vogue. By hindsight, of course, I think it's done a disastrous job in its presentation. It may sound like sour grapes—we're never written up in the press, because we just make money. But you see content companies that are going under written up all the time. And they perpetuate the myth that you can't make money with content. But you can. We've proved that you can. We don't get that press.

**CECILIA PAGKALINAWAN**　The building of the media exposure was great because it was a great story. Young people working hard making money—*the* American story. But when that story became young warped people, or young warped people on drugs, then it became, *We don't like these people anymore.* Initially we were the media darlings, and we became the media enemies because there were people who gave quotes against the media, and treated the media as if they were just hired help. There were also a select few people who just thrived on being in the media, and they still do. They're a small group, but because of the media they get they become representative of the rest of us, and that's not a fair comparison.

I remember sitting at a gathering at <kpe> for Hillary Clinton, and a few CEOs were invited. People were pushing Hillary on e-commerce tax breaks, and somebody from DoubleClick was there to push for her support on privacy issues, and I stood up and said, "Even if she put something like that in the ballot, nobody's going to approve it because nobody feels sorry for us. Have you ever sat anonymously on a bus or a train or in a restaurant? Anyone who's not in the dot-com world is hating us. They're talking about us negatively. They think there's excess here. They think young people are making all this money with very little work or no work whatsoever. Nobody likes us." Everybody just looked at me like I had two heads.

# THE NEW AD MEN

*T*he viability of content online seemed to depend on the ability to sell advertising on Web pages. Web companies eventually realized they'd need to find ways to support themselves.

In 1994, the first banner advertisements were served on HotWired and Pathfinder, establishing a precedent for advertising on the Web. Banner ads quickly proliferated, providing an essential source of revenue for content sites.

At first, Web sites tended to sell their own advertising in-house. But when an inventor named Kevin O'Connor hit upon a way to sell advertising on behalf of a network of client Websites, a powerful new force—DoubleClick— arrived in the Alley.

Buoyed by the proximity to major media buying outlets, DoubleClick was able to take first-mover advantage in the online advertising market. On February 27, 1998, O'Connor stood with his executive team and watched the stock trade successfully at seventeen dollars per share, then end that day at $26^3/_4$, ballooning DoubleClick into a half-billion-dollar company. It was the first significant Alley IPO.

*For the next twenty-two months, DoubleClick's stock rose steadily, rapidly, and repeatedly, splitting twice and ending a hugely successful 1999 with a closing trade on December 31, 1999, of $126.53. DoubleClick was a $15 billion company, and O'Connor a very wealthy man—at least on paper.*

*Flush with cash and valuable shares, the company went into hypergrowth mode, quickly becoming the Alley's biggest employer, with a worldwide work-force of approximately 2,000. DoubleClick maintained its status as the biggest and most successful homegrown Alley company through the end of the millennium. Its West Side office spanned six highly secured floors, and included a terraced outdoor basketball court on the twenty-third floor, a second terrace for cocktail parties and small events, a global map monitoring Internet traffic in a state-of-the-art command room, and ubiquitous free soft drink machines and coffee stations to help teams of engineers, technicians, and salespeople feel at home as they worked throughout the day and night.*

*Proud of its growth, DoubleClick hung a larger-than-life banner behind the Flatiron Building announcing:* DOUBLECLICK WELCOMES YOU TO SILICON ALLEY.

*Advertising played a significant role in the evolution of the Alley and the hype machine that made it rich. Offline advertisements were a key factor in creating IPO opportunities for companies as dot-coms realized they could grab just enough attention from an outrageous Super Bowl ad (like the single spot purchased by Richard Johnson of HotJobs.com with money he secured from a second mortgage on his house) to build a market for their stock on the day of its IPO.*

*The more cluttered the slew of dot-com advertisements became, the more surrealistic the ads needed to be to gain attention. Ad spending became an increasingly large share of a start-up's budget, and venture capitalists coached their portfolio companies to get big fast—or else.*

*That level of spending—especially for online advertising—far exceeded the ability of many of these start-ups to turn a profit. But through the end of the '90s, it helped build DoubleClick into the Alley's first Colossus.*

**KEVIN O'CONNOR** The first time I saw the Internet it was all government-supported, and so I thought, Someone's going to have to pay for the Internet. Our original product idea was to aggregate websites on a subscription basis; we thought subscription was going to be the model. But as we looked into it—in other media—we found that subscriptions would always lose out to advertising. So to us there was just no question that advertising was what was going to happen. People were going to fight it. I am sure in the early days of TV people didn't like the idea of an ad popping up; no one likes advertising. But someone has to pay for the medium. That is the reality. If you give a consumer the choice between free with advertising versus subscription, they always pick free. And just because someone wants to keep something free, that doesn't make it so. So we knew we would get resistance, but we knew that people would get over it.

**KEVIN RYAN**   We're an advertising infrastructure player. We sell Internet advertising, deliver Internet advertising, and provide direct marketing Internet services for Internet companies. We work with publishers and advertisers and try to make advertising work on the Internet.

**KEVIN O'CONNOR**   I grew up in Detroit; by the time I was about twelve I was kind of an inventor. I went to the University of Michigan and became an electrical engineer, graduated, and started a company right out of school in Cincinnati, Ohio.

My choice was to get a Ph.D. or go the entrepreneur route. This was 1983, the first year of the PC revolution, and I got totally addicted to that. I remember guys like Steve Jobs and Bill Gates, who were incredibly young—they were the first entrepreneurs of our time that anybody really knew. So I started a company that was tying PCs into mainframes, and we created the first product that tied a PC into a network remotely—the dial-in network. Of course we do that every day now, but that was pretty magical back then, around the mid-1980s.

That went pretty well; we were there for about nine years before we sold it to a company called DCA. In January '95 I started to really pursue the Internet. I spent the next eight months looking for ideas. I had moved down to Atlanta and was an angel investor, and I invested in a company called Internet Security Systems, which turned out to be a great one. We looked at a hundred different product ideas, and in September of '95 we started a company called Internet Advertising Network, which is what DoubleClick is today.

**KEVIN RYAN**   In the beginning we thought there might be three revenue sources for the Internet: advertising, commerce, and subscription fees. In the last five years, since subscription fees have fallen away, it's really advertising and e-commerce.

And if you try to decide now what's working better, look at the leading companies on the Internet. Many more of them are more advertising-supported than commerce-supported. So I think that advertising to date is working better, although they're really entwined, because if there's no commerce, you don't need any advertising, because there's no incentive to advertise.

Yahoo, Lycos, and C-Net are all advertising-supported models that are very profitable. There are many more examples, but if I asked you to name three profitable commerce-supported websites right now, you'd have trouble.

*What about cable television as one of the few subscriber models that work?*

**KEVIN O'CONNOR**   We actually did think it would look like the cable TV industry, where you get a premium package or a basic package. But the big difference with cable is that it's a total monopoly—you have no choice. You have a wire coming into your house. On the Internet you have infinite choices and so subscriptions, there is a fine line between total disaster and some success.

**KEVIN RYAN**   Before coming to DoubleClick I had launched the Dilbert site, which, in '95, was a precursor—one of the first sites on the Internet to be advertising-supported and profitable. So that was my introduction to Silicon Alley.

At DoubleClick I was employee number twenty. It was Kevin [O'Connor] and Dwight Merriman, who were in Atlanta at the time. What we realized at the time, which is still true today, was that there were thousands and thousands—in fact, tens, hundreds of thousands—of websites, and only a small, small, small fraction of those can sell advertising themselves. They really need to be grouped together to be able to sell advertising effectively.

It was clear that advertising would be sold in a fundamentally different way on the Internet from the way it is in print. Some firms just felt it was going to be the old model, but it's different—there's more of a direct-response element, and the targeting is even more important on the Internet than off.

You can do some things because of the technology that answer the question people in advertising have tried to answer for fifty years, which is, "What's working? What's not working?" Technology can help you solve that; that's what we've been doing for five years and are continuing to do.

The initial creation of the company was actually a piece of technology software. The fundamental development that changed Internet advertising was the concept of dynamic ad delivery. It seems obvious now, but at the time the fact that one user would come in at the same time as another and get a different ad—even though they're on the same site the exact same second—was what opened up all the possibilities of targeting.

What you require behind that is sophisticated algorithms to make sure that the right ad does go to the right person. You're counting it to make sure that we don't give too many ads for one advertiser and not enough for the other one. When you have more and more running at the same time, it becomes extremely complicated. When you think that every day we have to forecast how many people are going to come in during the next seven days from, say, San Diego, and how many people will be using Macintoshes, and how many people will be coming from the banking sector, and that some will be coming with Macintoshes in the banking sector from San Diego, you can imagine what it's like to make it all work at the end.

So that's the foundation. To deliver these ads dynamically, we have to have a huge distributed system so that we can make sure all over the world that we can deliver the right ad—very, very fast. Each ad is chosen in twelve milliseconds.

*What advertising was out there before DoubleClick?*

**KEVIN RYAN**   There was very little out there. First of all, when we launched advertising on Dilbert there were no models out there; we didn't know anyone who was doing it. All we did was put one ad up there and leave it up for three weeks. And since Netscape and Intel were the first two advertisers, and they paid us to do that, we thought we were doing great. Now, five years later, we probably have two hundred thousand campaigns running simultaneously, in an incredibly complicated model. And what's sort of fun is that Dilbert is part of our network. Because when I left the Dilbert site and came here, it came with us.

**KEVIN O'CONNOR**   We started the company and developed the technology, and then we ran across this company called Poppe Tyson, who had a group called DoubleClick. They had a concept similar to ours. They didn't have the technology, but they had four people who had been selling media, and so Dave Carlick of Poppe and I talked in early December. After a fifteen-minute phone conversation, we both realized we shared the same vision. We had the tech, they had the media. In January we brought the companies together.

It was very confusing. DoubleClick wasn't actually a company, it was a group of people sitting in Poppe Tyson, which was part of Bozell, which was part of BJK&E [Bozell, Jacobs, Kenyon & Eckhardt, Inc.], a big ad agency. Initial ownership was something like fifty-five percent BJK&E, forty-five percent us—Dwight Merriman and myself. That was big, going from being an independent company to being associated with a much bigger company. We picked the name DoubleClick because it was much better than Internet Advertising Network.

**NICK NYHAN** | FOUNDER, DYNAMIC LOGIC | Kevin O'Connor's office was in 40 West Twenty-third Street. I remember visiting him in his office—he shared it—and he was in one of those old boardrooms with wood paneling, and it looked completely silly. DoubleClick really got incubated inside Bozell and then Poppe Tyson, and it spun out. Because Kevin was smart enough to know that he wanted to work with every agency—rather than being purely a Bozell thing.

I remember once talking to Kevin when I was in Bozell—we were the internal research unit for Bozell—and he said, "Nick, we have all this data; we've got to do something with it. I want to know, at what point does click-through really drop off? Where does it plateau?" In other words, if you show someone an ad ten times, does it increase the chance they'll click on it? And where does it *stop* having an impact?

He gave us the server log data and we did the analysis. And he said, "If you get me this data, I can get it in the *Wall Street Journal*." Kevin knew he had a story to tell. I don't think he was arrogant; I think he knew that he was early and everyone was looking for answers. Sure enough, we found that it was three exposures—it was diminishing returns after three exposures. And that went in the paper.

**KEVIN RYAN**   Advertising initially emerged on the Internet, really, in 1995. DoubleClick launched in the beginning of '96. At that point the online advertising industry was probably about two hundred million dollars, from probably about fifty million in '95. To put it in perspective, today it's about seven billion. And I think it's been key for the development of Silicon Alley. Because, you know, the decision we had to face early on was, should the company be in San Francisco or New York? New York was chosen because it's the advertising center; a lot of the agencies and clients were here.

**DOUGLAS RUSHKOFF**   If you look at it from a purely capitalist perspective, what you want to do, in the limited number of eyeball hours you have, is maximize the number of purchases. If someone is online these days, it means they've come to make a purchase. If they've come to make a purchase, then a banner ad is distracting them from one purchase to try to push them to another.

The only exception to that rule is persons going to sites like NewYorkToday, or CNN.com, where they go to get information, and the site has to convert them from an information-gatherer into a consumer. NewYorkToday has to get you to make a reservation or to get a plane ticket; CNN has to get you to purchase an Amazon or BN.com book. But you don't make a conversion by taking someone off the track and showing them something different. You have to tease them or give them a certain amount of information about a story on a CNN thing, and then tempt them: Well, you seem to be interested in space. Don't you want to buy this new NASA book on space?

**KEVIN O'CONNOR**   Every medium needs a sort of standard unit. In the early years there were a hundred different sizes for online ads. Some were shaped like squares, some were shaped liked banners. It just so happened that that one banner was best for getting your message across. We tried vertical ones, but that only made sense if we were doing it in Chinese.

With DoubleClick, we created the first media rep firm. We were one of the first companies to sell advertising on the Web. We were massive. We used to sell Netscape, which was the number one media company for a long time [because it was the default home page for Netscape users, making it prime ad territory]. But they blew it. At one time we were like ten percent of Netscape's revenue, and it was an embarrassment to them because Netscape was by far the leading Internet software company, and software companies don't sell advertising.

On the technology side, I think that we brought a scientific approach to advertising, which flew in the face of Madison Avenue. We came from outside the industry, looked at the principle of advertising, and saw that it was actually pretty simple. Applying science to advertising is relatively impossible in traditional media, but it can be done on the Internet with technology. Advertising had been an art for so long that people really forgot the science.

**JASON CHERVOKAS**   DoubleClick arrived just as the attempt to standardize the web banner was being hashed out. Which I railed against editorially in the pages of @NY, because I felt we hadn't figured out what worked in this platform at all yet. And here we were ossifying this size.

Looking back, I might have been right in some ways, but I was wrong in others. And I didn't want to accept the arguments that were made to me then, but they were probably right—that media buyers needed to have a system of avails that they could understand. Production houses needed a standard, so they wouldn't be creating custom jobs every time. I think standardizing that stuff really opened up a flow of money that wouldn't have happened otherwise. But it was a double-edged sword, because it created this shape that doesn't really work as an expressive, communicative unit.

**NICK NYHAN**    My feeling is that banners are the bread and butter of the Internet. They are what they are. But from a common denominator standpoint, they are what you see most often. They're the easiest to implement, they're the easiest to deliver, easiest to track, and you know what? People expect them and are okay with them. When I see banners on top of a page, I know it's a trade-off, and I'm okay with that. Just like if I want to watch Riverdance on PBS, I know they're going to ask me for money. That's the deal. So I think banners are here to stay, and I think people who sound the death knell for banners are just trying to get attention. Banners will never die. They'll multiply.

**MARC SINGER**    A lot of people thought the whole idea of Internet advertising was silly. My little brother worked for Chiat/Day at the time, and he was like, "The Internet's not going to work. It's never going to be good for ads."

**KEVIN RYAN**    Most users—about seventy percent—come from a company, so they're at work when they're coming onto the Internet. If they come in from Citibank.com, Citibank is registered with Dun & Bradstreet as a financial services company, so we know they're coming in from that. But in such a situation you can't necessarily tell what department they're in or things like that. It's pretty precise, but not one hundred percent. If you come in from home, obviously we can't tell what industry you're from. We can just tell the geography. If you come in from Alabama, you'll show up as coming from Alabama.

There are data centers all over the world. Right here, we have an incredible data center—the network operation center. It's colossal; it looks like *Star Wars*. We've got cameras all over the place monitoring ad delivery all over the world, so we can tell instantly if there's a problem in, say, France. If all of a sudden ads are not delivering properly, we'll reroute from Germany to France, just like we'd do if we were air traffic controllers. There's about a hundred million dollars' worth of equipment in this building alone, mostly from the vendors. They're supporting this. And we serve ads from all over the world, so there are about twenty-one, twenty-two data centers throughout the world. If you're going to a French website, your ad will be delivered from a French server. And that increases the speed.

**NICK NYHAN** Recently I ran into a guy who, way back when, when I tried to get him into a meeting with Kevin O'Connor, spent five minutes in the meeting before saying, "Listen, this is great, but I have to go and meet another client." When I saw him the other day, he was like "What the fuck was I thinking?" Because he blew off Kevin O'Connor to go to some shmuck, and now he's like, "I am such an idiot." There was an attitude: "You guys are just playing, this isn't real money."

*DoubleClick ran into a stumbling block on its acquisition spree when it acquired Abacus Direct, a catalog database company, in 1999. The Abacus acquisition was decried by privacy advocates as a dangerous effort to merge online consumer information with offline addresses.*

**KEVIN RYAN** The Internet is new, and so from a combination of people's concerns and media attention, every issue that exists offline surfaces online and for a short period of time is made to seem unbelievably threatening.

Privacy is an important issue for everyone, offline. But over time people have come to the conclusion that, Well, maybe they get too many catalogs, but it doesn't feel like it's really threatening their life. I think online, in the beginning, people weren't really sure how this was going to play out. "What is really happening here?" And I think, over time, if you ask people, "What's the worst thing that's happened to you?" you'd probably hear, "Well, actually, nothing's really happened to me that's that bad." Over time you get to be more comfortable. What we've seen in our studies is that people who have been online for a year, and are using actively, are much less concerned than people who have just gotten online or aren't online at all.

A lot of it comes down even to the phrasing of it. People sometimes refer to DoubleClick as a tracking service. I don't think you'd refer to American Express as a tracking service for your purchases. Is AT&T just a tracking service of your phone calls? Well, no. Of course they have to keep track of your phone calls, because they have to bill you for it. But their job is just to allow you to make phone calls. And over time you're comfortable that that information is protected. We deliver ads in the way that FedEx delivers packages, and AT&T delivers phone calls. We have to keep a record, for billing purposes, that an ad got there. But we safeguard the information, and do a very good job on that.

The exciting—though occasionally frustrating—part of being a pioneer in an industry, and being the leader, is that you get the lion's share of the adulation as well as the criticism. I always cite eBay: things are advertised on eBay that are advertised in your average newspaper. But you can actually get a headline out of it that X sold on eBay, where if you say X sold on Kevin Ryan's website, no one really cares. It's the penalty of eBay being the leader, and we face the same thing.

*In late 2000, with DoubleClick's stock sagging under the weight of a slowing market and a loss of ad revenues, Kevin Ryan was named CEO and Kevin O'Connor stepped into the background as chairman.*

**KEVIN O'CONNOR**    Kevin Ryan has been here for so long, and everyone pretty much knew he was the one running the company. They didn't really see it as much of a change.

*i-Traffic, one of the first successful online media buying firms, was started by Scott Heiferman, less than two years out of college, in 1995. Everyone was trying to figure out how to make online advertising work—how to draw traffic to a site. Heiferman had learned some important lessons about online marketing during his first job, at Sony, in 1994.*

**SCOTT HEIFERMAN**    I was working at Sony; I was the first online guy there, in '94. I graduated from the University of Iowa, drove east because I'd landed a gig as an intern for this wacko marketing VP who had this wild idea that they wanted to think about online from a marketing perspective—which was just the kind of thing I was looking for.

Anyway, I was at Sony for about a year, and set up Sony's first Web presence and an area on AOL. AOL was the number three online service behind CompuServe and Prodigy. I set up the keyword "Sony" on AOL, which was really just a showroom for some Sony electronics products, some new PDA [personal digital assistant, e.g., Palm Pilot] type product they had at the time. And working with Sony, with AOL people, we launched it with great fanfare—keyword "Sony"—and no one showed up.

So now, trying to protect my own job, I started thinking about how people could actually find this keyword "Sony" on AOL. As anyone would have thought in my shoes, AOL has a welcome screen; they feature some links there; how could Sony be featured there every once in a while? Long story short, I went to AOL and they said they didn't even know how what's up there gets chosen. No one seemed to know. And I said, We'll write a big check. We've got something cool, a good offering, a neat little site on AOL. And they said, "No—we don't even have the facility to take your money, we're not a media company." Which is funny, because now they're arguably the world's biggest media company.

But it turned out that some friends I made in the process of making the site on AOL—some techies in the back room at AOL—were the ones who unofficially, day by day, hour by hour, made the decisions about what got put up on that welcome screen. So I sent them care packages of Sony Walkmans, and before you knew it Sony started popping up on that welcome screen all the time, and the traffic spiked every time we were on the welcome screen.

It was becoming clear that links would become an important currency, an important part of the online world, if the online world were to indeed take off. I wanted to focus on that—that whole question of how people navigate their way to stuff that interests them, how marketers get represented throughout an online environment. That was what was interesting to me. But Sony wouldn't let me focus on that, and I couldn't find a job focused on that. So I decided to quit my job, move to Astoria, Queens, live cheap, and screw around with this vague notion of an online advertising agency. And that was it.

*In addition to millions of investment dollars spent on advertising, entrepreneurs made themselves and their companies available as marketing*

*media by unleashing all manner of publicity gambits. From ubiquitous side-*
*walk graffiti and "snipes"—the posters that line construction sites—to wild,*
*gossip-column-worthy parties and the strip-down Nerve.com held to celebrate*
*the launch of its print mag, all pains were taken to attract public attention.*

**JOHN BORTHWICK**   We [TotalNY] put a guy named Greg [Elin]
on a motorcycle and put a little cam on his head, strapped to his helmet.
We put a modem on the back of his bike, and we sent him from Silicon
Alley to Silicon Valley, and along the way he met all the people who
were involved with creating the Internet. He met with Ray Tomlinson,
who invented the "@" sign and the protocol for the email structure. He
met with Tim Berners-Lee at MIT. He met with Skip Jones, who ran the
White House's Web stuff in those days. He went to ARPANET, and as
he drove across the country, every day he would post things on the
Web, including clips from his cam. He would have bikers who would
come follow him. It was about a three-week trip.

The website was amazing. We got tons of traffic, and we had good
sponsorship. We got Prodigy, AOL, and Sweet'n Low to sponsor us.
The first two, of course, you understand. Sweet'n Low was another
story. We found somebody who knew the president of Sweet'n Low. He
was fascinated by the Internet, and he said, "I'm going to sponsor this."
So he printed all these Sweet'n Low sugar packs with Silicon Alley/Sil-
icon Valley on the back. I still have a freaking box of them at home.
Everywhere Greg went, he gave out the sugar packs. One Man, One
Motorcycle, One Modem. October 18 through November 8, 1995. Silicon Alley/Silicon Valley.

People were like, This thing is so crazy, we're going to write about it.
And Greg was just a total star. So many people met him along the
way—so many of the media, actually—and asked him, "So, when are
you going to the headquarters of the Internet?" Remember, we're talking October '95. And he would say, "There is no headquarters to the
Internet." And they would say, "But it's got to be based somewhere."
And then he would explain that it was actually a distributed platform,
that there was no center. That, I think, blew a lot of people's minds.

We wanted this trip to have some kind of . . . civic purpose. We
wanted it to communicate something, to make a statement. It wasn't
just about goofy technology, although it certainly *was* goofy technology.

**JASON CHERVOKAS**  Mass media, as we know it, grew up in the first twenty years of the century, at the same time that mass retailing, in the form of contemporary department store and mass production and mass merchandising, were all growing up. And the state of the art in marketing then was to publish something like *Ladies' Home Journal* or the *Saturday Evening Post,* amass a lot of eyeballs, sell real estate for a marketing message that would drive foot traffic into stores to buy mass-merchandised product. That was the most efficient marketing machine that anyone had seen, and for forty or fifty years we refined it, and it was accepted as the model for financing media. But suddenly we find ourselves in an environment where marketing messages can get directly to the consumer, unmitigated by the media. I don't know that advertising per se is dead, I just think these relationships are changing.

**CRAIG KANARICK**  A lot of the work building the Razorfish image was done by word of mouth, but it was also done by us. Jeff's experience in marketing really helped out. We had a lot of contacts, business cards, a lot of people we met along the way. We also decided that every business card we got we'd put in the database; they started getting our faxes and our emails until they told us to stop. And we sent out press releases. We sent one out every week, which put us under pressure to do something newsworthy once a week. That prompted the famous fake press release that Jaime Levy wrote about Jeff tripping over his shoelaces.

But while people made fun of us for doing it, it also obviously worked. We wanted people to know what we were up to, and for every person that bitched or moaned, or thought it was fucking arrogant or ridiculous, somebody else said, "Add me to your list—I want to know what's going on," or "Great to hear from you," or "Wow, that's really great—congratulations." So it wasn't this oppressive thing we were doing to people.

We sent out a calendar our first year, showing how to take your old '95 calendar and convert it to a '96 calendar in five easy steps by rearranging the months and crossing them out and cutting off one column of Sunday and moving it over to the left. Now we do calendars every year. It was all part of building the image—proactive PR.

*Richard Johnson took a huge risk when he leveraged the bulk of his worldly possessions to buy a single ad during the 1999 Super Bowl for his company, HotJobs.com.*

**RICHARD JOHNSON** | CEO, HOTJOBS.COM | HotJobs started to do some offline advertising in early '98 or late '97 with Adopt-a-Highway programs. It was like four hundred bucks to adopt a highway, and we found these great spots—right in the Lincoln Tunnel and other major arteries through the country—and adopted them at incredibly cheap rates, because we were self-funded.

We were broke. We started doing job fairs so that we could advertise HotJobs and not pay for it. We figured if we raised $350,000 at one of these fairs, we could put $300,000 into local advertising.

Around May of '98, it became very apparent—maybe for the last five months it had been brewing in my mind—that HotJobs was not winning. Among the job sites, CareerMosaic was number one, Monster was number two or three, and we were like number ten, in terms of traffic. We started to realize that we had to be number one or two in order to succeed in this industry.

We needed a Hail Mary, basically.

We didn't have the money to run a Super Bowl ad, and I didn't have a VC, and we didn't have a way of getting VC dollars. So we went to borrow money from a bank, and we leveraged OTEC [the recruitment firm Johnson started before HotJobs], my house, my partner's house, all of our assets, and got a $4 million line of credit from Dime Savings Bank.

If I advertised in the Super Bowl, I felt, they would at least have to look at us before they discounted us and decided not to write about us. And we knew PR was very important. When we were selling our service, we were getting the question, "Well, where do you advertise?" Well, we advertised in the Super Bowl. It was something our sales force could leverage the whole year long. And there was the branding: People would see it and they would come to HotJobs and look for a job.

We realized we were only going to run this ad once. Everyone—the agency we were working with, our PR firm—told us not to release it before the game; it would be like giving away the punch line before you tell the joke. You just didn't do it. But no one ever came up with a better reason than that. So I released the commercial ten days early. On television, it was played in its entirety thirty times. It was played in bits and pieces a hundred seventy times, totally free. We had Boston, San Francisco; two DJs there talked about Fox rejecting our first commercial, one of them for forty-five minutes.

The AP came to our commercial and shot us as we were filming. The photos were sent over the AP, and on the Friday before the Super Bowl, it was on the front page of a hundred and thirty-five newspapers.

Basically, we became the story of the Super Bowl. If you went from Tuesday to Friday before the Super Bowl, we were covered more than the Denver Broncos. I mean we had eight hundred media hits, eleven national news programs. *The Wall Street Journal* picked it up, ran a story on it. The press just blew into our doors. At one point in our old office, we had ABC News on the sixteenth floor and the NBC News on another; the two stations were at our office at the same time waiting to interview us. It was an unbelievable avalanche of media coverage.

We went from being a small Internet company to a brand overnight. That was the catalyst that made HotJobs.

After the Super Bowl, we couldn't even afford to run the ad again. But our traffic went from three hundred thousand in December to about five, six hundred thousand. And then in January it was 1.6 million; in February 1.7 million. We were bracing for the slide back to three hundred thousand, but in March it was 1.6; in April it was 1.9, so it actually tracked down and then back up.

People always ask me: "You mortgaged your house—what would you have done if all that had failed? What would have happened to HotJobs?" And I turn around and say, "Well, you would never have heard of HotJobs."

*The success of the Super Bowl spot propelled HotJobs into an IPO seven months later. Johnson kept up his commitment to advertising, and long after the market crash in 2000, HotJobs.com ads still adorned subway tunnels and buses across New York City.*

**ANNA WHEATLEY** They decided that they were marketers. I have no direct evidence, but it became a kind of one-upmanship, and the

Super Bowl was the ultimate end goal. Whether it's viral marketing or the first-to-market fallacy—which is a total fallacy, bad business any way you look at it, and not true—they believed it. They thought, Let's spend a lot of marketing dollars. And CEOs started getting all this coverage overnight for nothing. Then they had to come back to earth. Marketing—branding—is not simply a matter of putting up a billboard in Times Square or running an ad during the Super Bowl. It has a purpose, and that is customer acquisition and retention. That's it—it doesn't sound very sexy, but branding is not a means in and of itself. That was a very big learning curve, and the VC community itself got swept away from fairly sound business principles that they probably already knew.

**DOUGLAS RUSHKOFF**   Around '98 and '99, the Silicon Alley companies admitted to the public that they themselves were meaningless by migrating to television in thirty-second ads about who they were and what they did. When Internet companies realized that the way to look cool was by buying television, was when they really started to burst the bubble. That's when it became clear that they don't mean anything. Unless you advertise in the Super Bowl, you don't exist. So the relationship was odd.

For two or three years, it was as if traditional media companies thought the Internet was trying to replace them. They were trying to figure out, *How can we absorb it before it absorbs us?* Meanwhile the Internet, which knew it had about a hundred times more money invested in it than it was actually worth, needed to use that capital then and there to buy some traditional media companies that actually had some value before the stock price collapsed.

And new media conducted a successful enough public relations campaign through old media to get old media to believe that new media was worth something. So Time Warner let itself get bought by AOL [in 2000]. When AOL did that, it was obvious and clear: The Internet was cashing in its chips.

**JOHN YOUNG**   I think a lot of dot-coms were essentially advertising to Wall Street when they were buying Super Bowl commercials,

They weren't really focused on bringing in customers; they were trying to get the biggest bang for the buck (which is television), to get the biggest amount of awareness at once. If you're trying to do something that's going to drive the most site traffic and bring in shoppers, online is probably the best. I think that was the folly of last year. Everyone went on a feeding frenzy, and all the television properties were sold out, because everyone was just going, Oh my God, we have to make this big splash. And how do you brand? Television. Which, being on the interactive side, we just thought was very ironic. If you're an online medium and you're trying to build your business—if you're trying to reach customers—you should be doing it online. And during the last two quarters we've seen a backlash against that. Everyone's switched their TV budgets to online budgets. And they're now being more responsible to their stockholders.

Essentially, the way you build a brand is not through one medium. There are all these different points of touch with the consumer—what you see on television, what you see in print, what your friends tell you, what you see online—and you have to be a master of coordinating all of those and have an agency to help. I think a lot of dot-coms were under pressure from their investors. You need the big splash; we need that big sizzle for Wall Street, to drive up the stock value. They were focused on Wall Street, not customers. You look at where they're placing a lot of those ads—shit, a lot of those ads were going on Sunday morning. That's where most of the investors are going.

A lot of people building dot-coms were looking for the fast buck. They were looking to build their business and sell and get out. Built to sell or built to last? It's obvious in their advertising strategy. A company like Amazon is trying to take the approach of built to last. They're being cautious about how they build.

**JASON CHERVOKAS**  I did an analysis a couple of years ago that was pretty interesting, comparing About.com and Yahoo. At the time, About had just reached the same traffic measurement levels Yahoo had reached during the quarter when it first turned a profit. But About was nowhere near turning a profit, and it was only about two years later. And the principal difference was that at the time Yahoo was spending something like forty percent of revenue on marketing, which was a high number but reasonable for a start-up growth company. About was spending something like a hundred and twenty-five percent revenue on marketing.

As much as anything, it was just a measurement of the increased noise level you had to differentiate yourself against. But I think it pointed to why the guys who had gotten there early enough had actually managed to make this hybrid model work, and sort of con us all into believing that this was in fact the right path, when it was really just a moment in time for which it was the right path. And we still haven't found the right path.

**CECILIA PAGKALINAWAN**   It became overinflated, to begin with. It was just not logical at all. Having worked with brands and clients who are very traditional, none of them ever spent more than fifteen percent of their money on advertising and marketing. There's a certain formula to it, and there's a reason for the formula. All of a sudden, you started hearing that people were spending seventy percent of their overall budgets on marketing, and there was no way you could sustain that. It was irrational exuberance. It was definitely hyped, hyped by everyone—the media, the PR agencies, by the companies themselves. Eventually that hype was going to burst.

**ALAN MECKLER**   I fortunately have always been able to see three or four years ahead, and the Internet today is not what it will be in the future. In the Internet of the future, brand will mean nothing. I have a saying: The greater the number of bus posters, the more likely the company will go bankrupt. You can probably figure out who those companies are, but they will all go under, because the brand means nothing in the Internet.

Intelligent agents and bots will find all the information that anybody wants, the best information for the cheapest price. That's why Internet.com is bulletproof: If you have the best information, the bot will find you.

But most of the models we see in New York City today, and around the world, are based on the assumption that brand is important, that you should spend a lot of money to build your brand and it doesn't matter if you don't make money. I'm suggesting that is a total waste of money and will be meaningless by the end of 2001.

*It was the plan of almost every content and community site that advertising would sustain it and allow it to grow rapidly. Community sites, which sought to aggregate users by subject, interest, gender, race, or by serving as a place where users could read content, make homepages, chat with like-minded users, own a free email address, and communicate with one another, were enjoyed by users en masse. Community was an extremely popular concept early on, especially after GeoCities was sold to Yahoo for five billion dollars in 1999, setting a benchmark entrepreneurs could always rely on when presenting themselves to investors. Huge companies based on different concepts of community— including iVillage and TheGlobe—were supported by investors looking for similar performance.*

*Some have said that the idea of the community site is a fundamentally flawed one, that the Internet is best suited for two-way communication services like Instant Messenger and email. A more likely explanation for the downfall of many—though not all—of the community sites by the end of 2000 was the inability to aggregate enough users to be an effective media brand. With the Web so easily catering to a huge number of competing properties, few websites attracted the size of user base that made television, until the advent of cable, such big business.*

**OMAR WASOW** Community is this totally overused buzzword. I still use it because I know what I'm talking about, but I'm never sure that what I'm describing as community makes much sense to other folks. What I mean by community is technologies that allow people to communicate with each other and to express themselves. GeoCities, for example, was a way for people to express themselves with personal Web publishing tools, but it wasn't really about communication. If you were a movie fan and you set up a site in the Hollywood section and somebody else set up a website in the Hollywood section, you didn't necessarily ever talk to any of those other people. So was it community? It was community in the sense that it was user-generated content; it was self-expression. But it wasn't a very rich form of content, because it wasn't about interaction between people with shared interests.

There were many attempts that were unsuccessful. "We've got a magazine, let's put up a chat room. We've got *Car Lover* magazine, let's put up the *Car Lover* chat room." Well, if you've only got ten people an hour on your site, that's not a chat room that's ever going to get critical mass. It's not going to take off. So they'd say, "We tried community; it failed. End of story." Well, it's not enough to have a chat room, and it's certainly not enough to have a chat room that's poorly maintained and doesn't get much attention.

You saw a lot of people dip their toe in community, or dabble in community, but not produce the kind of robust, fully realized version we've tried to do at Black Planet by providing people with lots of ways to communicate, so you can send somebody a note—an instant message; you can chat with them; you can send them an email, which is slightly different from a note; you can look at their personal page; you can sign their guest book. All these different ways to communicate enrich the dialogue; having just one tool, on the other hand, is like only being able to send postcards.

AOL is the best example. Before AOL, if you wanted to chat you might have used one tool: You had an IRC [Internet Relay Chat] client, you had an email client, you had a Usenet client; you had all these different pieces of software designed to let you use a chat room, a message board, an email. What AOL did was put all of that together.

**CLAY SHIRKY**  The idea was, if communities are successful in that they get people interested, then someday it will be possible to charge money for them. But the number of communities you are willing to pay money to belong to is actually very small. Maybe a gym. Maybe a club of some sort. If you're a kid, maybe the Boy Scouts or Girl Scouts. But there are just not that many communities you'll pay for.

One of the unfortunate truths of the media that no one has gotten past is that people like to get recommendations from a different source than they buy things from. This is a normal part of media. Some categories are exceptions: You're perfectly willing to talk to a book clerk about what book your Aunt Mary might like. But you're going to take car advice from someone other than the person selling you the car.

The notion that communities would be these hotbeds of commercial transactions proved to be false. The advertising industry switched from brand building to more of a direct-marketing focus, and the click-through became the normal measurement of sucess. But what community does is produce sites where people are interested in staying right there and talking to their friends. Chat boards generate page views, and therefore attract banner ads, like crazy. But even in the old days, click-throughs were abysmal. Back in the day of the two percent click-through, a chat page was generating sub half a percent. Now that sub half a percent is the norm for a regular page. God only knows what it's like for the chat sites.

**DAVID LIU** The advertising model has fallen out of favor with investors, because people have started to question whether there is going to be the growth people expected. I think the idea of the "tech-wreck" is indicative of the loss of faith in the investment community, and an erosion of faith for certain business models. The VC money has stopped; the wanton spending has stopped.

From our standpoint, this is not dissimilar to '98, when there was a busy signal problem at AOL. It was front-page news in the *International Herald Tribune*. AOL was jammed and couldn't answer the calls. And we turned that into a PR coup for ourselves; we went out and said, Yeah, people can't get there, but guess what, our site is swamped! People can't dial in because people want it. It's not a bad thing—this is a great thing. We just have to catch up to the demand.

With this advertising scare people are suddenly saying, "There is the death of the CPM. People want quantifiable results." Well, at The Knot we've been selling ads to endemic bridal-category people. Some of them don't even have fax machines, let alone a computer. The first ad we sold was to somebody like Nicole Miller. And they were like, Great! Where do we get a banner? Where does the banner go? And then you realize that you have to educate them on the entire process.

For a lot of these people, their measurement isn't some sophisticate at Ogilvy & Mather saying, "Well, you have twenty-five thousand presses on this, we'll spread your media dollars like this." They are saying, How many times did my phone ring this month? If this went online, did I make any sales? They want quantifiable results. Our feet have been held to the fire for the last four years in the most painful way, because if it didn't show results they didn't come back and advertise. The fact that everyone now is being held to the same fire—"welcome, you're four years too late"—I don't shed a tear for those who complain about that.

It's also interesting because we battle in an environment with *Brides* magazine and *Modern Bride*, who have just strained a level of trust and credibility. They give the magazines away for free at the expos and then say, my magazine has a twenty-six X pass-along rate because it sits at Dottie's Hairdressers, or it goes to XYZ Bridal Salon, and you combine all these to get a circulation rate that's unrealistically high, and hard to believe. There are 2.3 million marriages every year. Only 1.8 million actually have weddings. Who are these magical people you're selling these inflated CPMs to?

When we started delving into this, we said, Oh my God, there are so many business practices going on here that really raise questions. I think that has happened in other categories as well. The Internet has added a level of accountability and measurability that's going to completely change the way people deal with advertising. The winner in all of this is going to be the consumer, and those people that deliver true, tangible results.

*How does the consumer become the winner?*

**DAVID LIU**   If an ad program doesn't work, that means that it isn't being looked at or responded to by the consumer. Consumers will only respond to things that are relevant to themselves. The better I get at engaging my consumer with relevant information, the better I am serving my consumer. It's kind of like what Nicholas Negroponte once said. At some point advertising will become news. Personalization information on each consumer is so well captured and capitalized upon that it will be news. There are some organizations that are actually getting very close to it. People like Amazon are doing a phenomenal job. I bought a book from them, and they sent an email to me yesterday that said, We know you acquired this book, you might be interested in this other one because it talks about some of the same things. I was like "Wow! That's great." And I bought it. That was news—not a coupon that came in my Sunday circular. It was a message targeted for me, to see if I wanted to preorder. That is what advertising will eventually have to become, because online is all about delivering a result through the advertiser.

**DOUGLAS RUSHKOFF** The World Wide Web itself is an electronic strip mall. The user as customer is already in the store when he's online. Any content is actually an obstruction between the customer and the buy button. So advertising doesn't work in that space because it's a distraction from buying one thing into another. Really, the way to make the World Wide Web work the way Silicon Alley would like it to is to get out of the way, and turn e-commerce itself into the entertainment experience—to make the actual clicking on the button, the buying of the product, as effortless and seamless as possible. Once the person's logged on you've already made the sale. You've isolated them, which is what mainstream media has tried to do to people over half a century. So they're alone now, and they're in a place where the only way they can participate is by making a purchase. Anything else is beside the point.

# RAISING MONEY IS A HUSTLE

*Venture capital became the lifeblood of the Internet start-up. By making a high-risk investment in the early stage of a new company, venture capitalists hoped to recoup their commitment many times over when the company IPO'd or was acquired for a huge lump sum.*

*In the early days of the Alley, there was almost no New York–based venture capital to be had for Internet start-ups. But that all changed with the creation of Flatiron Partners, the first major VC fund set up specifically to fund early-stage technology start-ups in New York City. Established by Jerry Colonna and Fred Wilson with money from Chase Capital Partners and Softbank, it became one of the real leaders of the community.*

*As Alley companies began to have successful IPOs in 1998 and more institutions saw golden opportunity in early-stage investments, a few dozen funds were established for new companies.*

**JASON CHERVOKAS**   By the 1970s, New York was dominated by investment banks and leveraged-buyout shops. The type of risk capital we associate with Silicon Valley just didn't exist as a culture in New York. There just was no apparatus for investing in early-stage businesses of any type.

**ANNA WHEATLEY**  The West Coast had a lot of technology investment because it had thirty years to develop. You cannot change the physics of time. California had produced many generations of entrepreneurs, and they all knew each other and knew how to work together, and they were all primarily from a technology manufacturing background. East Coast money was tied up in huge, big-money deals. The private equity group at Salomon Smith Barney was looking at a million-dollar, two-million-dollar, even a five-million-dollar investment, and they're like, "I've got a hundred million dollars." They are used to doing huge-scale deals; they were used to dealing with those Fortune 50 companies, the Fords and Hearsts of the world, and doing huge mergers and acquisitions. West Coast money knew how to do incremental investing.

**ANDREW RASIEJ**  The old-money VCs in New York, which are tied to traditional businesses, got established through investment banks and other banking institutions. They had processes in place for judging potential investments. There wasn't a real strong network of seed or angel funders in New York, and the old-school VCs were really focused on funding manufacturing businesses or service businesses, where you could really see what the market was. So the only ones doing dot-com funding were either those in a position to take a huge risk, or not-very-experienced VCs, who tend to be trend followers. Now some of them actually claim to be experts.

**JOHN BORTHWICK**  The first round of capital [for TotalNY], we raised around seventy-five thousand dollars. That was in November or December of '94. The key investor was a brain surgeon who was a friend of a friend of mine, and who thought what we were doing was fascinating. I characterized him as a dream capitalist. He's like an angel investor, but he was investing in just a total dream. He really believed in the dream, and he was willing to put his money on the table.

People used to say, "You really have a brain surgeon who is your investor?"

Everybody else who became part of the company also put in a little bit of money, like two or three grand.

Some people invested their computer. They owned their own computer, and they capitalized that, and that became their investment for some equity.

**DAVID LIU** Raising money in New York for The Knot was an impossibility; people just didn't get our concept at the time. Our second round of financing we secured through Hummer Winblad, one the premier VC shops in Silicon Valley. But we were probably their first business-to-consumer commerce or media-related East Coast investment. This was from left field. These were people used to investing in accounting packages and switchers and things, and we were talking about bridal gowns and registry products. But Ann Winblad is a real visionary. And I also have to say that, because she is a woman, she really got it. She sat there and said, "Oh my God, people would just die for this," and she just jumped on board. Even then, people were very dubious about the growth of online advertising.

**JON EFFRON** | FORMER SALES ASSOCIATE, EGROUPS | Mike Moritz, who was our investor, came to the office for the day once to speak to us. He was our interim CEO for maybe a month or two, but he was basically being a super VC and trying to recruit a CEO. We expected he would talk to us for five minutes. There were ten of us there and he talked to us for forty-five minutes.

At a certain point he started referring to his investments, and how he had found [Yahoo founders] David Filo and Jerry Yang in a trailer in Santa Clara, and how they had a great idea and someone needed to back them and this and that. He went through the whole story of Yahoo, and after giving us the riveting turns about stock splits and valuations, he just stopped and paused and said, "It's interesting. Sometimes I forget how many zeroes there are in a billion."

And he just started laughing, and no one knew what to do. It's not that funny a joke, but the fact that someone would make it and be able to back it up was like, *Whoa.*

**SCOTT KURNIT**  We went through three rounds of venture financing before we went public. And we started with a loan before that, so it was really four rounds of financing. People talk about the difficulties today, in 2000, but back in '96 and '97 there were cold spells that were quite scary.

I think each round of financing is a turning point. Our company has now had five or six, if you include the first loan. Two rounds of public financing and four private financings. Each one of those is nip and tuck. It's how much of the company are you going to give away to someone who's going to fund the company—and is there even someone there to fund the company? Fortunately for us there was always someone there. But you tighten your belt in anticipation of cutting those deals. I'm sure my wife remembers the couple of times I covered payroll back when we were sixty-one people. It's either your money or no money; you do whatever it takes.

**SYL TANG | PRESIDENT, HIPGUIDE.COM |** It took me a very long time to understand that if I had a man in the room with me when I went to a business meeting, it went better. He could sit there like a bump on a log; he could be my brother, or someone I recruited off the street, and it would still be an advantage.

**FERNANDO ESPUELAS**  I had been working at AT&T Latin America, and about a year after being there, when I was twenty-nine, I was promoted to managing director. I had focused all my energy on making it big in the company.

Then I had this Groucho Marx experience: I didn't really want to be part of a club that would have me as a member. I felt idiotic, because I had spent so much time and energy to get to that point, and I thought what I needed was a little perspective.

My wife and I decided to go to India and Nepal, to get out of context so we could get some context. In Nepal itself I had this life-transforming experience, a moment of clarity like I've never had before. It was simply that the Internet was not just convenient, but really represented a historical shift as profound as the arrival of the Spanish and the Portuguese in Latin America. If we could get it into people's hands, we could change history.

The problem was, I was on a mountain. I couldn't do anything with the idea. But I started to write in my journal ideas of what I wanted to do.

When I got home I called my best friend since sixth grade, Jack Chen, and told him my idea. He was just enthralled by it. Even though he wasn't sure what the business would be, he was excited that *I* was so excited. That's when we started working on a plan, which eventually became StarMedia.

**JASON CHERVOKAS**   Silicon Valley–type venture capital grew out of the PC industry, and the PC industry is a much different business from the Internet business. I think those guys didn't understand that. They missed out on a lot of investment opportunities because they were thinking, Well, we have a patented chip architecture that we can understand, we can productize—we can fly an IPO up the flagpole, and it's a business. But this was a very different kind of business. There was a disconnect between the type of risk capital that existed on the West Coast and the type of businesses that existed here.

**JACK HIDARY**   The road show [where you pitch investors] is an intense experience. I highly recommend it—it's one of the most wonderful experiences I've ever had, and I've done it three times now.

You have seven meetings a day for ten days. On top of that you have a lunch and sometimes a dinner where you speak to many investors at once, but the seven meetings are all one-on-one. So you meet with seventy, eighty, ninety investors. That's a lot of investors. You tell the same story again and again, and you get the same questions again and again. Every time you've got to make it fresh. If you enjoy telling a story, that's okay. If you don't, it's not such a good thing to do.

You start on the West Coast first, then you go east. You go to San Francisco, L.A., San Diego, maybe Portland. Then you move your way east to Kansas City, Chicago. Two key areas there. Strong Funds is in the middle somewhere. You go to Michigan, Alabama. Then you keep moving east. You get yourself to Florida, the Tampa/St. Petersburg area, Boston, and then you end up in New York. Oh, sorry, I missed Texas—Dallas and Houston are important. It could be up to four cities in a day, in some cases. The only way you can do this is with a private plane. Bankers arrange for it. It's the only way to get around for these road trips. Otherwise you just can't make it. You just can't make it. Every CIO, CEO I spoke to agrees. So usually two, three cities a day, sometimes four.

**ANNA WHEATLEY** We decided to take New York City [entrepreneurs] on a road show to San Francisco, back when every investor on the West Coast said that they weren't interested. It was a very hard conference, but we just *knew.* I remember talking to a group of employees here in town, and they all said it was naïveté, I didn't come from venture capital. All the West Coast VCs, meanwhile, were saying, "Why should *we* get on a plane?" Well, why do you think New Yorkers have been getting on planes investing in West Coast technology companies? It's because you have an obligation: You have money, and it's not your money—it's a pension fund's money. And your job is to go wherever there are great deals. And it's saturated: There are going to be more markets, other than just the West Coast, that are doing technology—but on the application and the content development side, not the manufacturing side. We very much believe in that model—your job as a venture capitalist is to find a good deal.

**FERNANDO ESPUELAS** We put together our savings and went to Staples and bought two chairs, two phones, two lamps, two tables. And we were able to borrow from Jack's family these three rooms in an old house that they were about to rip down to build office space. We said, "We'll be here three months and then we're out," but we were there for over a year. The rent was good—it was free.

After we put out the table, I laid out every business card I had ever collected. I said, "In forty-eight hours we'll have our first deal," and I never quite understood until that moment what momentum is all about. When you start a company you have no momentum, and every single moment you have to create your own. I called everyone, and then waited, and expected for things to happen magically, and no one called me back because I wasn't at AT&T and didn't have a forty-million-dollar budget anymore. So the beginning was very, very exciting—but I realized that if I were to get a cold the company wouldn't exist that day.

**JASON CHERVOKAS**   It wasn't until the launch of Flatiron Partners that people began to wonder: Here we were sitting in the financial capital of the world; how could we create an apparatus for early-stage financing with these companies?

**FRED WILSON**   I think what we saw in 1996 was the opportunity to create a venture capital firm that was focused on the Internet here in New York City. Looking back on it, it seems like that was pretty obvious. But at the time it was *not* obvious, and in fact when we announced that we were going to do this, people would say, What, you're going to be local—*and* you're going to focus on a single industry? That's never going to work.

**JERRY COLONNA**   In fact, that's exactly what they said in the articles.

**FRED WILSON**   Softbank had been an investor in a number of deals with me, and they were looking to try to get some feet on the street in New York, so they had proposed that I join with them and cover New York for them. That was intriguing to me, and I was thinking of maybe doing that, but also thinking about starting a venture capital firm—and that's when I got hooked up with Jerry. We went to see Chase and convinced them to match Softbank, and put the two seventy-five-million-dollar commitments together. And then we went out and started making investments.

Jerry and I started talking in the spring of '96. We got the commitments for the money in the summer, and we made the first investment in August.

But the industry, the Internet? Turned out it wasn't an industry, it was like a big megatrend that impacted lots of different industries. So we had the opportunity to invest in media, retailing, technology, services, telephony, and telecommunications, all these different things. The one single theme was the Internet. So it became more of a marketing thing, sort of a rallying cry, than a limiting strategy.

**JERRY COLONNA**   It was amazing, because the week we were putting everything together—including our first investment, which was Yoyodyne—the Java Fund got announced out in California, and the *Journal* had broken the story on [Kleiner Perkins'] Java Fund. Through what was essentially a slip-up, the *New York Times* found out about our fund. They ran with a very large story, like a page one business section story; that evening we were on CNN. From the beginning we sort of stepped into this media vacuum that existed, and it was a sort of a roller coaster, a rocket ship ride, from the beginning. Because the amount of press and media interest in our story was way out of proportion with what we were doing at the time.

**FRED WILSON**   'Cause we were doing nothing.

**JERRY COLONNA**   Right. We'd done one investment, and we very quickly became spokespeople for the whole investment strategy in Silicon Alley. And I think it took a couple of major investments and some time before our track record began to match our press. But that's very typical of New York. It's true about baseball, any other kind of industry. When you're here in the media capital, you make some noise and it has reverberations around the country. And that's really what happened with us.

**FERNANDO ESPUELAS**   We were looking for cash all the time. The dentist of one of our employees was our big target, and I think he invested fifteen thousand dollars, which was amazing for us.

I had eighteen credit cards. As I was leaving AT&T, I'd said, "Well, you never know," and I started filling out preapproved cards. I got so many at the same time; the computer didn't know I had taken them all out. We would take money from one credit card to pay salaries, from another to pay the phone bill.

*What was the investment landscape like in New York City in 1996?*

**FRED WILSON**   There were a lot of service companies. Razorfish, Agency, SiteSpecific . . . and the funny thing about that was, we never thought those companies could scale to be big businesses. We were wrong about that. We eventually made an investment in iXL, so we did actually get onto that bandwagon. But in the beginning we weren't all that intrigued. And then there were a number of e-marketing companies—Yoyodyne was one, Bigfoot was one, FreeRide. There were a lot of people thinking about the Internet as a marketing medium. There was DoubleClick, which we missed. They got funded just before we got started.

**JERRY COLONNA**    iVillage also got funded just before we started. In fact, we cited iVillage being funded by California-based VC firms as an example of the need in New York for VC firms. There were a lot of very young folks who really didn't understand the venture capital process, the fund-raising process. I would say that the vast majority of companies we rejected for funding we rejected because they just weren't appropriate. They were looking for fifty thousand dollars just to get started, and we were looking to put a million or two million dollars at a time.

*How would you explain the process of raising venture capital?*

**JERRY COLONNA**    Typically, an entrepreneur will raise capital from friends and family; that's known as an "angel round" or a "friends-and-family round." Or maybe they'll put in capital themselves to get the business started.

While it's not always the case, the first-time entrepreneur is better served by launching a business, getting it to generate some revenue, and maybe even breaking even on an ongoing basis. At that point they can present a good case for a two- to three-million-dollar investment that catalyzes the business and takes it to a whole new level of revenue, potentially even profit. And at that point they should really be talking to an institutional venture capitalist. The distinctions between angel-level investment, seed-level investment, first-stage venture capital investing, later stages in venture capital—those distinctions were lost on people in this area in 1996.

Today, there are enough publications out there, enough dialogue, that people can understand that process. But it is a very step-function process. People can short-circuit that, but generally the people who are most successful in short-circuiting that are those who have raised capital before.

So we have a CEO whom we've backed before and made a lot of money with, and he's gone on in life and started a new business and needs to raise capital; if he calls us, he doesn't have to go through those first rounds. He generally can raise additional money. If you're a rock star—Marc Andreessen raises fifty to a hundred million first-round investment, and that's a reflection of the fact that he's perceived to be a serial entrepreneur and a success.

*How were you introduced to StarMedia?*

**FRED WILSON**  Jerry got a call from a woman named Anne Andiorio, who is in the PR business. Jerry used to be the editor of *Information Week* magazine, and Anne used to call him with story ideas. So there was a relationship there. Fernando and Jack, the two founders of StarMedia, had hired Anne early on to help them get the buzz out about StarMedia. And so really the whole company, the first time we met them, was Jack, Fernando, a creative director, a sales guy, a couple of programmers, and Anne Andiorio. Anne called up Jerry and said, "I want you to look at this business plan." And it was actually a pretty well done business plan for a company that was eight people. The reason was that Jack used to be at Goldman Sachs in corporate finance, and he knew how to write a business plan. They had a pretty simple value proposition—to create an Internet portal for Spanish and Portuguese speakers. That seemed smart—Jerry thought it was smart anyway.

**FERNANDO ESPUELAS**  We could not convince anybody on the planet that the intersection of the Internet and Latin America provided one of the most important opportunities in the Internet revolution itself. Today Latin America is the fastest-growing market in the world. When we spoke to Latin Americanists, they did not believe in the Internet; they did not understand it. They were still talking about commodities and steel. And when we talked to technology people even they didn't understand the Internet.

One really big VC here in New York—after my big pitch, his total answer was, Latin America is that big blob south of Texas. That was very representative of people's point of view.

**JERRY COLONNA**  Fernando's business plan sat in the slush pile of unsolicited business plans for a good two or three weeks before I finally got around to reading it.

**FRED WILSON**  And he read it and came into my office and said, "You know, this is kind of an interesting idea." Because the idea that . . . you're not going to create another Yahoo in English, but the Spanish and Portuguese speaker isn't necessarily going to go to Yahoo. So is there an opportunity there? And this was in the Spring of '97.

**JERRY COLONNA**  It was a year after Yahoo's IPO.

**FERNANDO ESPUELAS**  We had someone at the company who was doing the networking, and we tried Flatiron several times; at that time Flatiron was the hottest VC, and we knew they were going to focus on the Internet in New York, and we couldn't get it to mesh. Finally, after an enormous amount of persistence, they gave us a meeting. And it was the kind of meeting they do very often: You've got half an hour—prove yourself. The meeting went fantastically well. It was with Jerry Colonna and a couple of people they brought in, and he was fascinated. Everything he heard made sense, except he couldn't put it together with his knowledge of Latin America because he had none. *If* what I said was right, this could be very big, but he didn't know if it was. So, although it felt very good, we didn't know.

**FRED WILSON**  And so we said, This is an interesting idea. I remember we went to see them for a second meeting, in this little tiny house in Riverside, Connecticut.

**FERNANDO ESPUELAS**  They had said, We want to come in and visit you, and we said okay. They came out to Riverside, and Fred Wilson—I always made fun of him because he was so rude. It was a bit of a dump—or, rather, a total dump—where we were. So it was very difficult to imagine that this would become what it is. He walked in and looked around, and—he didn't spit, because I guess he's too polite for spitting, but it wasn't far from that.

We had a great conversation, where he kept challenging everything we said. Then we took them to the back to do a product demo. Of course it didn't work. Our music video channel crashed at that very moment. But they understood that it was the people, rather than the product, that existed at that point.

**JERRY COLONNA** The funniest part was walking in there. The reception area was a desk and two folding chairs, and I remember the folding chairs were right next to each other, and we had to sit like this [scrunched next to each other] while we were waiting to see them. We're like, Who are they kidding? We're just waiting here. And Fernando said, "Please, step into my office," which was the chair over there.

**FERNANDO ESPUELAS** They liked us enough to invite us back to this massive meeting, where they brought in the Chase people, who have enormous experience in Latin America. It was at this meeting that we explained our vision, explained where we wanted to go. It was a very interesting meeting, because when I would say how Latin America would take to this, the Latin American people from Chase would say, No, no, it'll be *faster*. They immediately understood that this was going to be big. That was a first, after forty-some-odd meetings with venture capitalists. We were out of cash; it was a pretty desperate moment. It was the first time we met that intersection between technology and Latin America.

Until then, a common question was, Do they have phones or computers in Latin America? Which was somewhat mystifying. Another question, which was asked of me about six months ago live on CNBC, was, "What's the point of what you're doing since now there's going to be translation software?" I've even had people ask, "Don't Latins like to talk face-to-face?"

**FRED WILSON**    It was a very unimpressive operation. But the presentation they gave us was classic Fernando, just perfect. They had it nailed in terms of what they wanted to do, and the vision was right; they really were convincing that there was an opportunity.

**FERNANDO ESPUELAS**    I believed that the division of Latin America into multiple countries was purely an accident of time and space. Once you removed the power of the Spanish crown or the Portuguese crown, what you had was this collapse of central authority. What's been forgotten is that Latin America was unified for over three hundred years and it's been divided for fewer than two hundred. But over the last two hundred years we've all been taught to think that people on the other side of the river, or people across the mountain, were somewhat alien beings. The fact is that Uruguay, where I'm from, and Mexico, have never been part of a contiguous whole—but we speak the same language [and] have a cultural connection; a lot of the cultural context is the same even though there's enormous amounts of difference.

That historical reality, combined with the fact that sixty-five percent of the population of Latin America is under the age of thirty—in the two largest countries, Mexico and Brazil, half of the population is under the age of twenty—I believe that this huge new baby boomer generation in Latin America, growing up with the Internet, would have a completely different concept of the world. Especially when given the opportunity to connect with each other for the first time in two hundred years, without the filters and without the oligarch telling us why the other people were aliens. That would connect people in a new way, and could lead to a huge historical transformation for the region. And that's exactly what's happening.

**JERRY COLONNA**    Part of the presentation that Fernando made to us that was evocative was when he would stand up and talk about reuniting Latin America and overthrowing the shackles of colonialism, the artificial boundaries that were overlaid on the continent. And you could sort of sit there and think *this is kind of crazy*, but that kind of passion is infectious, and it's compelling, and it drives companies to create really interesting things. The fact is, if Fernando could unite Latin

America, and have a company worth [only] a hundred million dollars, he'd consider it a success.

**FRED WILSON**  So we went back, and we noodled on it, and we really liked the two of them and thought they would make a great founding team for a company. We did a lot of homework, enlisted the help of the Chase Latin America private equity team. They gave us a whole perspective on Latin America, what the size of the company could be, all that stuff. And we ultimately decided to do it. We bought a third of the business for three million dollars.

**FERNANDO ESPUELAS**  We were supposed to close the first round of financing on a Friday at two P.M., and Fred had asked me to appear at six or seven with Jason from *Silicon Alley Reporter*—he had a show on Pseudo, and we were supposed to close and do the show and talk about it. It didn't close, but we went on the show anyway, and of course we're running out of money, and I was thinking *It's all a game.* Of course you get really paranoid—it's just the surreal nature of talking about this three and a half million dollars that we don't have, feeling that we're now part of the group. We didn't close the financing until about 2:30 in the morning, because it literally takes that long. But at three or four in the morning we got the sense that everything had shifted—we had made this enormous transition from an idea that could have died to a company with a future and a lot of responsibility.

**ANNA WHEATLEY**  Jerry Colonna came out of a magazine background, but he had also worked his way through the business angle. And I think that he is a very smart guy. I understand his position much better than I do with Fred Wilson. Fred is much more the money guy, the traditional guy. When I picture Fred I think of someone who salivated over making the best deal—you know what I mean? Jerry just kind of followed the flow, and he wound up together with Fred doing Flatiron. But I have never heard any complaints from any of his companies. I think that he is overworked, but really genuinely trying to think—What can I do? What can I do to help this company?

**ALICE RODD O'ROURKE** It took quite a while for the money to start flowing in. One of the reasons the East Coast VCs started funding was that the West Coast VCs started funding, and territoriality was an extremely important social phenomena. Local VCs said, "I'm as smart, I'm as good, I want to make the same returns as those guys." What is this that's going on in my own backyard? By 1998 the activity was beginning to pick up during the second half of the year, and then there was a total explosion in 1999. I also think that there was an amazing, world-class, first-ever model that had been created out on the West Coast. You not only had people looking over at their VC colleagues, you had people looking at the West Coast, look at how well it worked out there, thinking maybe *we* should be part of that.

**ANNA WHEATLEY** Suddenly you had early-stage people from Kleiner Perkins at iVillage or Hummer Winblad at The Knot—some very big names—trying the borders here. The East Coast money didn't have to get on a plane, and that was a very attractive proposition to those who were doing the early-stage deals. By and large, the New York VC firms didn't warm to the idea until they saw that they could make money in the market, and that was all they needed: Overnight, everybody had a seed fund. All of a sudden they could all afford that five million dollars.

**KYLE SHANNON** In late 1995, there was money, but if you didn't have something that was in a shrink-wrapped box, they wouldn't give you any of it. It started to change early in '96. We started getting calls from VCs who didn't understand what we were. They would sort of come in and say, "We'd like to invest in you. We always did want to run ourselves one of them Internet companies."

Chan [Suh] and I were not too fond of turning over the keys to some money yutz who didn't really have any interest in the business other than as a cool hobby. And then we met Omnicom. And Omnicom came in and said [in Russian accent], "We're bigger than God. We could kill you if we wanted to, but we would like to work together." [He smiles.] It wasn't at all like that; it was a good meeting.

We decided to accept a forty percent investment from them; we wanted a minority investment, which meant that Chan and I still retained control of the company, still had the board. But it gave us access to their bank.

*Bob Lessin, CEO of Wit Capital and former vice chairman of Salomon Smith Barney, also became a major investor in early stage companies.*

**ANNA WHEATLEY**  When Bob Lessin was at Salomon Smith Barney, before he joined Wit, he invested in forty different companies, twenty million total, and over the course of one year he became man of the year for everybody—for *Crain's*, for *AlleyCat*, you name it. It was a brilliant strategy. He was the top investor for New York, meaning he had the best deal flow. To this day there is no one who doesn't send a plan over to the Bob Lessin camp. Forty companies—that's a strategic move. You're hedging your bets. Who has really kind of missed the boat? I'll tell you who hasn't done particularly well—the bigger corporate and media investors. They were all looking in New York early, and I don't know that they have had any big successes.

**DAVID LIU**  Even after four years, four rounds of financing, and going public, the most difficult decision we made at TheKnot.com—and the decision that best represents what it means to be a dot-com in the city—was the decision to move into our new space. In February or March of 1998, we were in an office space the size of a small conference room. We had eleven people churning away, and we were gaining some momentum selling ads, and we had just locked down a three-book deal with Random House, our first step toward multiple media. We were never comfortable with being a dot-com or a website. We were always The Knot; we had great aspirations to become this great big media company. We aspired to be Martha Stewart, as opposed to TheGlobe or iVillage or something.

We'd been hustling to create a mock-up of a new magazine that we could show around. Hachette was interested, and asked how quickly we could staff up. So here we're panicking, because we can't staff a magazine in this space—we were going to die. We decided we needed eight other people, which as I think back on it was just absurd. Someone told us about a loft that was subletting space, but it was a little ominous because it was a CD-ROM company that had gone public a year before and had gone out of business within a year. They said, If you want the whole space you can take it. So we talked to the landlord and he said, Just give me your financials.

This was around February or March, and we were going through money rapidly, and we had a couple of engagements but nothing that we were really sure of. So the guy looks at our financials and says, "What the hell? You guys are going to be out of money in a month! I can't rent this space to you." We were like, "No, no, no, we are about to close this deal, but we can't tell you who it's with." He said, "Look, you're going to have to put down at least six months' security." I knew how he calculated six months, because he looked at the balance and knew that that was exactly how much money we had. We had literally only forty thousand dollars left.

So we sat there and debated, and it was getting toward the end of March. Someone else wanted the space April 1, so we had to make a decision. Here we were, sitting in heated negotiations for financing with an understanding that we were going to try to get it from Hummer or Hearst, but we didn't know which one was actually going to step up. I had to decide if we were going to spend our last dollar on ten thousand square feet when we only had ten people. How bullish were we on our prospects? The weekend before April first I said, You know what, if we aren't going to be bullish on our future now, then what's the point? It was a bit reckless: we thought, If we go down, we want to go down in flames. If we don't close the financing, what is that forty thousand going to matter? It would buy maybe a month and a half. It wasn't as if we could resurrect the financing; we would have to call it quits. We had to jump in with both feet. My partners and I certainly didn't want to divvy up the forty thousand and say that's it—that was not what we were in this for. So we said, "Let's do it."

It was a little absurd. We didn't even have enough money left to hire movers, so we were rolling our computers down the block on our office chairs and into the elevators. And it was cavernous here. It was ten people in ten thousand square feet. We moved in and we closed financing twenty-six days later. And within nine months we had run out of space.

**RUFUS GRISCOM** What's been extraordinary in the last couple of years is watching the VC community like a little herd of llamas go from one business model to the next. When we first launched it was all about search engines. "Are you going to start a search engine? Are you going to create a sex search engine?" Well, we did create a sex search engine, a directory. Then it was all about community, homepage building; then all about e-commerce; then it was all about B2B [business-to-business] and wireless. It seemed to change about every six months. It's fantastic to see the investment community realize that they've been running around following all these sorts of very faddish trends, and kind of wake up to that.

**ALAN MECKLER**   Flatiron has obviously been a leader, because of some of their early investments that did well. Although God forbid if we tally them up today to see where they are. You know, that's going to be a big problem. Some investors are going to have a big problem going forward.

**KYLE SHANNON**   At the time we did the deal with Omnicom we were profitable, we were making money, we were fine, we were growing. But what we got was that this industry was now picking up momentum at such a speed that we weren't going to be fine if we didn't start growing more quickly. So we used the deal with Omnicom to finance some of our first acquisitions, and actually some of our second, third, fourth, and fifth. And it's been a really good partnership.

I would say it changed around 1996, and the Gold Rush probably started in '97 going into '98. That transition makes me a little sad. Because prior to that, the only reason people were quitting their day jobs and starting businesses was because they passionately believed in something. And after that, no one really gave a shit if the business was good or not—no one was passionate. It's like, "Oh he's doing a travel thing? I could do a travel thing—fuck it. Give it a different name, get a new URL, IPO baby, yeah, whoo."

There was a party mentality, like a video game group had invaded our book-reading club. *You don't gotta read—we can kill each other! Whoo!* That was the dynamic.

It felt like a very trusted, quiet community that was generating a lot of power. A *lot* of power. It was really smart people, very willing to share success and failure so that we could collectively learn more quickly.

If there was a mission about WWWAC, it was that: to demystify this whole thing that everyone was trying to get into as quickly as possible. Because the theory was, if we can demystify it for everyone, yeah, someone might give up a trade secret or so, but it's moving so fast that if the whole boat goes up dramatically quickly rather than one or two going up, then we'll all be better off.

But by the end of '97, early 1998, that spirit started to die. It was all money, all bottom line. "Get in a manager. The wacky kid with the vision? Get him out of here." So many companies got acquired or invested in, and the investment bankers took over and said, "Okay, kid, great, thanks. No, seriously, get out of here. Clearly, you've made something here. Now leave."

**ANDREW RASIEJ** When I was trying to fund Digital Club Network, we did get seed-funded by a venture capitalist in New York. A guy by the name of Dillon Cohen from a company called Carlin Ventures funded us a seed round, and when we were going around trying to get our second round of financing, we were thwarted by what I call GeoCities-itis. When we were looking for funding, GeoCities had sold itself to Yahoo [for a whopping $5 billion] and so all the venture capitalists and a small group of people who all know each other made a lot of money on GeoCities [including Flatiron Partners]. And then every deal that they did from that point forward had to look, smell, and taste like GeoCities.

That lasted a couple of months, but I got the sense that there were a lot of venture capitalists who were lucky enough to be involved in this area of funding, as opposed to housing or other businesses—manufacturing or otherwise—but then really didn't have the experience or the breadth to take a long view, to really analyze where the potential was. And I think you're seeing now that a lot of them are paying the price, because a lot of companies they funded . . . I mean, there's a lot of bad business plans out there.

**ESTHER DYSON** New York investors tend to be more out of the financial community, and the ones in California tend to be more out of the high-tech community. What you really need are people with domain experience. People with some business experience—that is, who understand the business of actually producing some product or service. A lot of people say, "I'm going to go redo such-and-such industry and create an exchange," but they don't know much about how that industry actually works.

I tend to invest in people who do. So I've recently made investments in somebody in a logistics company whose father grew up in the shipping business, and who also did himself before he went to work for UPS in logistics. And somebody in a company in recruiting who began as an offline recruiter, and somebody in receivables management that was running a debt collection agency for eight years, so he knows about collecting receivables. That's the kind of experience you often find lacking both here and in Silicon Valley. Here they think it's all a question of finance and managing a business, and there they think it's all a question of technology and creating the software.

Often it's a question of really understanding the business. You need to understand something before you can change it. The challenge, of course, is that if you understand it too well you might not be able to see how to transform it.

**SYL TANG**    I was on a plane one day before my site launched, back when I didn't know who the big names were in the Internet. I started chatting with a guy in his fifties. I didn't know who he was, and I think he was really interested in me, because I didn't care, I didn't know, and I didn't want anything from him. I was probably the only person he'd met in the last fifteen years who didn't want a single thing from him.

He was on the board of a number of Alley companies—some that are still around and some that are not. Once I realized he was somebody in the industry, with a lot of interesting things to say, I wanted to know him. He was someone who had been there, done that, and I could learn a lot from somebody like that. But we were not interested in the same thing. Over a number of lunches it became really obvious to me that he was interested in something more, and I was not. I was never unclear about it.

We'd been to lunch and a couple of dinners where it should have been clear that it was a business dinner between acquaintances, friends, people in the industry. But then he started inviting me on trips he was taking. He made it pretty clear that he was interested in me, and I made it pretty clear that that was not where we were going.

Soon after, I went to a launch party for a friend's site. The guy in question was a minor investor in their company, so he was there. He was just starting a fund, and he had a business partner, and he introduced us. The business partner had clearly never heard of me, and I said, "It's nice to meet you," and he asked what I did. After a while our mutual friend walked away, so his business partner and I were left making small talk. So I told him what I do, and he said, "Oh, you should talk to this company we're in the process of acquiring—we just gave them two million dollars."

I asked what it was they do, and he said, "They publish a wireless guide with cool information on places to go and things to do."

I tilted my head to the side and said, "Really? What is that URL?"

I frantically made a note in my Palm Pilot, ran to the bathroom, called up our chief programmer from my cell phone, and I told him to get to the Web and check this site out. Then I went back into the party, and I didn't know what to do, so I decided to confront this guy I'd been lunching with. I went up to him and said, "I'm heading out. I'd love to have lunch with you next week."

He looked off into the distance and said, "Yeah, I might be out of town. I'm not really sure."

I said, "Tell me about XYZ company."

And he just looked like a deer in the headlights. By that point I had known him for six months, and I had never seen him startled. But he was startled. Because he was caught completely off guard.

He said, "How do you know about it?"

I said his partner had told me.

He said, "He shouldn't have told you about that."

I said, "Are they a competitor of mine?"

And he said, "No, you'll beat the pants off of them."

I said, "If that were true, you wouldn't have given them two million dollars!"

I had never gone to him for funding. We had never talked about it; it just wasn't right. I would never take money from someone who wanted to date me. It's just a conflict of interest.

He turned to my friend and said, "Make sure she's picking up the check at dinner, because she's going to be a very wealthy woman someday."

I was just floored. Then he turned back to me and said, "This wouldn't have happened if we were on a more intimate basis."

*Venture capital investments and stock prices ballooned in tandem from mid-1998 to early 2000. But when the NASDAQ began its precipitous decline in April 2000, a long shadow was cast over the success of Alley VCs.*

**JASON CHERVOKAS** The venture capitalists buy in at different stages, and they're locked up in different ways in terms of when they can sell. I think typically what happens is: they invest; the company goes public; then, over the course of its life as a public company, they sell parts of it, or they don't even sell it; it sits in the fund; and they borrow against the capital gains. Insiders—both founders and early-stage investors—are typically locked up for a certain period of time, typically six months, nine months, sixteen months, whatever. So they don't necessarily sell on day one. Most of the secondary offerings you saw during the heyday of the Internet boom were insiders selling, and that was often first-round investors and entrepreneurs, selling at least a part of their stake as soon as their lockups expired, to take some money out. Which is fine; I don't think anyone begrudges anyone making money. The question is, Were they just cashing out because there wasn't a business there and they knew it?

*Most Silicon Alley companies anticipated a fairly long lead time to profitability, but the agencies that formed in the early days were often profitable from the beginning. When the market crashed in 2000, the agencies that had taken on substantial financing and gone public were discharging so much money to cover their burgeoning businesses that when the amount of spending by companies on new Web products tapered, they were forced to fire workers and scramble to reduce their burn rates—the rate at which they were burning money each month.*

**ANNA WHEATLEY**   The VC community underwent its own crash course education. You had the creation of the celebrity entrepreneur—and you know what? The VCs were loving it. They love to see their guys on CNN. They love to see them on a cover of a magazine. So the company gets kudos for that kind of publicity. I think that the VCs have always taken on the mantle of business development: Use the Rolodex to attract the talent, make the deals, get you in with someone you can make some deals with.

*When did things become more competitive?*

**KYLE SHANNON**   Everything became so much more competitive in late '97, early '98. It was like a light switch, because when the money came . . .

When the money came, it came on a lot of fronts. Clients started spending a lot more money, so we got dramatically busier; VCs started coming in, throwing money around. So when someone gives you ten million dollars, you all of a sudden go close your door and start working, and if someone gives you ten million dollars they probably made you sign some pretty scary documents that say If you fucking tell anyone about this we'll come back and fucking rip your head off and smack you around with it, got it? Right. So that probably killed a bit of that collaborative spirit. And then the whole IPO market. To go IPO there's a long run-up where you're legally required not to say anything. And as that stuff intermingled together, it really just sort of severed the ties. All these really strong, trusted ties just got severed very quickly as this stuff came into place.

And then, what followed that was that a whole group of people came in who didn't have the same motives. Listen, everyone wanted to make money. We weren't stupid. But we were going to make money because of this vision we had, not because *there's this thing, and we can make money with it*. And that's the second group that came in. Once they came in, some of these guys intermingled with them, and there were intermarriages and stuff like that, and all of a sudden you couldn't really tell people apart anymore. And now it's not a trusting environment. Now it's a sharky environment. It just is.

And it's sad, but I think that those early days really were responsible for the fact that there's venture here at all. A lot of money came in based on that early pod—blob—of optimism and pure invention. Pure collaborative invention. And then in '97, '98, it changed, and we all came out into the world.

**ALAN MECKLER** I think what's going to happen in the next year or two is that a lot of so-called smart people in venture capital in New York City won't look so smart. I would say that by the end of 2001, eighty-five to ninety-five percent of all the stars of venture capital in New York City will look like dunces. Mainly because they invested in ideas that were already passé, but they didn't realize it. Because of the fool's gold of the stock market, the press built them up to be really smart people, geniuses. Some of the genius people—I won't mention any names—put a lot of money into Inside.com. You had to be a fool to invest in that—certainly based on the original business model. Maybe it will evolve. There are a whole lot of public companies and not-so-public companies, and venture capitalists that don't really understand the Internet. [Inside.com could be called a competitor to Meckler's Internet.com.]

**ANDREW RASIEJ** In the end, the things that would drive anybody's investment in manufacturing or services are the same things that would drive investment in dot-coms: Do the people who manage the company know what they're doing? Do they have vision and passion? Is there a market? What's the execution potential? How big is that market? How much money is needed? Those things don't change, whether it's a new media business or an old business. But there was a period of time where, just 'cause a company had "dot-com" associated with it, everybody said, "Okay, great." Those days are over. And, in fact, I'm very happy that we had a shakeout; everybody woke up and said, "Hey, maybe this isn't so great." It's good, because it forces a lot of these bad business plans off the market. It allows for new employees, good employees, to find better jobs with companies that have potentially better business plans, and it gets people to focus on the bottom line— what's real, rather than what's perceived.

Unfortunately—and I think it's only natural—Silicon Alley just went through its hype period, and now it's reality-check time. But a lot of people made money along the way.

**RUFUS GRISCOM**   The first few meetings I had with investment bankers, I showed up in a full pin-striped suit and a starched shirt. But it became clear to me almost immediately, from the expressions on their faces, that they hoped I would be dressed differently. Because I represented for them an alternative lifestyle, and for a lot of these guys on Wall Street a form of employment they would prefer to have. So to see me looking like themselves was kind of disappointing.

*What's the worst pitch you've ever heard?*

**FRED WILSON**   It might be the one for the service where you could order fast food from your car and pick it up when you get to a toll booth.

# THE NEW WORKER

*B*y 1999, Alley Web companies were brimming with energy, funding, and business.

The explosion in capital on Wall Street fed all manner of Internet companies. Agencies could now play to two markets: start-ups powered by zealous venture capital and a mandate to build, build, build, and old-economy companies with boom money to spend on their Web ventures and digital strategies.

The Alley exploded with jobs, employing as many as two hundred and fifty thousand workers by the end of the year—a far cry from the enlightened handful who made up the generations of '94 and '95.

The Alley spread out physically across Metropolitan New York, and there were now more than eight thousand companies in it—nearly a third producing annual revenues of at least one million dollars. Together, the companies had raised nearly six billion dollars from investors and the public markets.

In New York there was money everywhere, and knowledge workers from a dozen industries were tempted by analogous online jobs: online editor, producer, business development associate. If the new job didn't pay well, then it was probably for a pre-IPO start-up—that great species of potential money-

*maker that could pay out millions of dollars at odds far superior to the lottery or casino.*

*The quality of working life in the Alley seemed profoundly different from that of jobs outside the industry—at least on the surface. Most of the entrepreneurs who had built their companies from scratch placed emphasis on a less hierarchical, less strictly managed organizational system. Making hay while the sun shined, many workers put in sixty or more hours per week, turning the office into a home away from home. Amenities made it easier: a basketball court and free soda machines at DoubleClick, plenty of free bagels, pizza, and snacks elsewhere. Open floor plans exposed rows and rows of wired desks, and workers liberated from their cubicles traded in their privacy for a view of the windows.*

*The communication tools available to them made the workday even more intense, with silent staffs communicating entirely via Instant Messenger, and high-speed access to rich media entertainment, Pseudo programs, or Internet porn, giving them a new way to take mental breaks during the long day.*

*Most important, workers were armed with stock options—shares of the company to be purchased in the future at a significantly reduced rate. The bond made them owners in their companies, and as owners they found that their relationship to the work changed forever . . . or at least as long as the stock was still rising.*

**MARISA BOWE**    Before the Web, people like me didn't know what the fuck we were going to do. It was hard to figure out. Okay, I have my degree in semiotics, and I wrote my masters thesis on Herman Melville. Now what do I do? I don't have any skills.

All these people were stuck in grad school, never finishing, or heading off to law school or business school with deep resignation.

Now everyone can get these pretty good-paying jobs doing stuff that isn't too horrible in an environment full of people like them.

**CLAY SHIRKY**    I started in the revolutionary camp. I'd come from the not-for-profit theater, and I thought capitalism was this world of bloodless calculation. When I saw how business was actually run, I thought, This is chaos—we're just making it up as we go along. Then I realized, Everybody's doing that. The point of capitalism is that no business plan fully survives contact with the market. You get your ideas, you take them out, and then the market destroys some of your assumptions in an absolutely merciless way.

There was a group of people who said, essentially, "I don't want to know from that. Don't talk to me about the market; I'm here pursuing my art." And there was another group of us who said, "Well, if that's how it is, let's get it on." That second group of people became management—not necessarily because we were any more qualified in training, but just because we weren't trying to preserve the illusion of being a cultural institution.

**NICHOLAS BUTTERWORTH**   Like many people of my generation or my age or my tendencies, I never planned to have a corporate job or be a businessperson per se. I just wanted to do interesting and intellectually challenging things that I thought had some social value. The things I had done—playing in a band, political consulting, journalism, and Rock the Vote—were all very different experiences, in a very compressed time frame, but that's just the kind of people we were. We weren't looking for one stable situation. We were looking for lots of diverse things. Part of the whole Gen X feeling was that there wasn't going to be any loyalty to any one company—one company wouldn't be loyal to its employees. And there's going to be this fluid work environment anyway, so why try even to pursue one career? Why not just do different things?

So when the Internet opportunity came up, I think there were a lot of people who had been knocking around the media business, not necessarily on a career path per se, but interested and engaged in it, and this was just the perfect opportunity for them. We were open, available, and looking for something.

**THERESA DUNCAN**   Some of the people I started with thought they were really, really going to the top—that it was going to be this evolution where they would just get higher and higher.

But it was very difficult for a lot of people. Some came up against brick walls and just started spinning their wheels. You think you're going to be vice president, or that everybody's going to have stock options, and then you're a producer or project manager for three, four, five years. You keep putting in the same hours and working on the same kinds of projects, and that gets disappointing for some people.

A lot of people just veered into doing advertising and things like that. They were just churning out websites for pet food companies, but they thought they were going to be David Geffen, that they were going to take the world by storm. But it just wasn't the case.

**CLAY SHIRKY**  That these businesses were more socially conscious or more flexible or fluid or whatever was largely a function of the pioneer mentality: "We don't care where you are, we don't care what your past was like, we're not going to look at your résumé. We're not going to call your references, because they don't understand what we're about to hire you to do anyway. And if you can look me in the eye and say you can do this job, let's do it."

And so we ended up with this motley assembly of people—almost necessarily people who had nothing better to do with their time—and it gave the industry a really unusual flavor in the early days.

**SCOTT HEIFERMAN**  In '95, when you left a six-figure job to come work for i-Traffic in my apartment, you weren't getting paid much, and it was kind of a smack across the head. You knew that you had to question things, take a different approach to work, throw away any notions of what an office was.

**KEVIN O'CONNOR**  We were very careful when we screened people. We really wanted people who wanted to change the world. The chance of getting to pioneer a new media is really almost nonexistent. New media come along every fifty years. And so we tested people, getting them to take pay cuts when they came in, getting them to change their lifestyle when they came in, and to take a lot at risk.

**KYLE SHANNON**  The Web has made the line between work and play extremely blurry. And when you're working in an environment where you're always using or developing these technologies, even the line between taking a break and working is blurry.

In designing our office space, one of the things we realized is that sometimes ideas emerge when you're on a break. So we've got white-boards and blackboards all over the place, and spontaneous meeting rooms you don't have to book out—you can just run in a room and close the door and write from floor to ceiling on the whiteboard.

That blurring between work and life is very much a reality, and I think it comes out of the fact that this is still so new. Five years ago no one knew how to do this; there was no training for it. They were doing it in their evening hours, until two in the morning. They'd go to work with bloodshot eyes, and people would ask, "Were you out drinking?" "No, I was up coding HTML." All those people quit their day jobs and started this industry. So the stuff they were passionate about was now their job. Which was really cool—but now it's a blessing and a curse, because it stops being a hobby after a while, and stops being as enjoyable.

**STEPHAN PATERNOT** | COFOUNDER, THEGLOBE.COM | Every-one now, in this industry and everywhere else in the world, is educated about stock—that you've got to make lots of it. But back then we didn't even know what stock was.

When Todd and I started the company we hired a law firm in Cali-fornia, and they pretty much told us we had to create a company struc-ture. We knew nothing about business. Even though we just wanted to do the cool Internet thing, we had to learn the entire business part. As we learned it all, we discovered that everything's structured in terms of shares and options. You could pay people cash or give them the stock, but the reality was, "What the hell is this stock?"

We would talk to people about it but they didn't know what it was. We'd try to give it to them because, hell, it meant we didn't have to pay as much. But they would say, "No, no, no. I'll take this stock, but give me $6.50 an hour instead."

**TODD KRIZELMAN**   Today people will walk in and they'll say: "Is it a nonincentive stock? Is it an unstatutory option? What are the vest-ing terms?" It definitely did change the feeling of running our company.

**RUFUS GRISCOM**   All these people spent years ignoring their friends' kids, or some bumpkin intern who was yapping about the Internet and the implications for their business. Now that haunts them,

and they walk around with a tic, like a nervous twitch—*never again.*
Young interns offering vice presidents advice are listened to much
more attentively than they were seven years ago.

Entry-level salaries in the book publishing world just went up to
thirty thousand dollars, from twenty-four thousand, and that's because
people are losing their best talent to Internet companies. It's been cool
and empowering all around. I think the real shake-up in the media
industry is just beginning.

The next ten years are going to be outrageous. Disney and Time
Warner and all these large players are really going to have to fight and
move fast on their feet to try to keep their claim, because the landscape
is changing so radically. I just think it's cool.

To me, it's less about the Internet as a religion than as a process of
breaking up and complicating the power that controls media. I don't
care how it gets to you.

**DOUGLAS RUSHKOFF**  In the early Internet time it used to be
that employers wanted to create a cool space in order to attract cool people. Now the cool space and the cool people are the advertisement for
their business. If you want to bring someone in and show off: We've got
pierced kids working here. Look, there's a skateboard leaning against
that desk. The kids themselves are for show as much as for work.

**JON EFFRON**  The image that will always stay in my mind was
sawhorse tables with thin Sony Vaio laptops in the Internet office where
I worked. They were pulling up by their bootstraps, so they had
sawhorses, but they had Vaios—which are terrible computers, but they
look really high-tech.

One of the guys interviewing me for the job showed me the list of people who had invested in them—the same guys who invested in Yahoo,
Cisco, and Apple. He gave me ten seconds to digest it, and he said, "If this
doesn't work out, you can work at any of these companies. Once you start
and get in at an early level with one of their investors, they'll always be
there for you. It's kind of like the Mafia. You decide to sacrifice and take
the risk, and they'll be there for you." Whether or not that's true—when I
knew that next Monday I'd have to get back to my cubicle at Disney and
fill out some requests for proposals and do real low, menial work—to

have someone say that a VC investor would look out for me if I took the risk to join their company, and I'd get my own Sony Vaio and a cell phone, I said, That's a no-brainer, I'll take it. I didn't need to see a salary.

**NICHOLAS BUTTERWORTH**   When we first moved in, we had only enough employees to fill up a third of the floor. Then there was some open space, and my office, separated not with walls, but by a buffer zone.

Just as we were moving in, someone said we should have a basketball hoop. Not as a gimmick to recruit, but because we had too much space and were bored. So we started playing a lot of horse and one-on-ones. We had the over-the-sprinkler shot, the off-the-cyclorama shot, the off-the-column shot, the bounced-off-the-column-off-the-ceiling—all the good shots. It was great, because when musicians would come over they would shoot with us.

**KEVIN O'CONNOR**   Our belief has always been that, from day one, every employee has options. Everyone here had ownership, and was treated with dignity and respect. In my first company we didn't give options to everyone, and I was against that. I always wanted to give options to everyone. When everyone is an owner, I think people look at things differently. But early on, getting options in New York was very new, and people looked at an option agreement and thought, "Is this a scam? I'm being obligated to buy stock?" It was a strange experience. People didn't value that, so we took a lot of unique approaches. We offered people more money or options—whatever they wanted.

**ANNA WHEATLEY**   Now we know that this is a new economy, and we're in a transition, and we're all fortunate enough to be in it already. But the really exciting part is going to be how we grow the businesses, and grow the workforces that make the businesses work. And New York is just not a natural habitat in some respects for that. It's very independent, very creative, but it's kind of that turn-of-the-century image that I have of all those workers building the Brooklyn Bridge, building the subways. It's a different kind of labor force, but it is a labor force. We're going to see that manifest. Now is the exciting part. The other part was just to get us there and open the gate.

**ESTHER DYSON** Workers are more sophisticated in general. It's not just stock options. They have more choice, and the flow of information the Internet fosters really has changed the balance of power between workers and employees. But it's not because they're working at start-ups, it's because they have more information.

Workers are getting a greater share of the equity, because they're more valuable. It's not good news for the investors, because they keep getting diluted. Capital is worth less and people are worth more. But I think that's great.

**CLAY SHIRKY** I've had a ten-year run in a business that was nothing but uncertainty. In the theater, one of the things you learn is that you're always looking for a job, even if you already have a job. People who come in without that baseline assumption can certainly feel betrayed by a system that let them down, in that the business wasn't as normal as it presented itself. On the other hand, all Internet businesses are sort of crazy, and they're all trying to kind of back themselves into a normal way of doing business.

If you're running a thirty-person company, it's really kind of a tribe. Pretty much anyone there really knows the score, one way or another. They certainly aren't privy to everything, but you can just get a vibe. Everybody sits in one big room.

I remember one afternoon at SiteSpecific, looking out at the pit, just standing there with this gorgeous afternoon sun streaming in. The place was dead quiet, but the concentration was fierce. And I remember saying, "This is SiteSpecific firing on all cylinders." Everybody was focused. You didn't have to look at anybody's desk; you could just feel it.

When you're working in a thousand-person company, you have no idea what it's like in the Singapore office. You barely know what it's like down the hall from you.

**ANNA WHEATLEY** The mania enticed people, especially those in their thirties or forties who had topped out where they were, or had seen the top and found it wasn't that interesting. It goes back to the spouse or significant-other issue—the other people in your life whom you have to convince—because very few people have no responsibility to anyone. All of the sudden it got much easier to say, "This is where we are, I've saved up this much," and convince your family and your friends that this was worth doing. It was liberating. It was like, "Oh my God—now I don't have to be stuck as an investment banker for the rest of my life." Or "I don't have to be an account manager on Madison Avenue. I can actually do something on my own and create something and be part of this vibrant community." That is still here. That is totally the upswing.

*Has the advent of the Internet introduced new ways of working?*

**DOUGLAS RUSHKOFF**  These kinds of work situations, apparently, have popped up throughout history. Even Socrates was talking about a group of ten or twelve men speaking their minds honestly. Chiat/Day, in the 1960s, created these playgrounds for people. All these Algonquin Round Table things. This egalitarian management style peeks out every once in a while.

The interesting thing about it is that it almost always involves a charismatic at the top—a cult-leader type guy. And I look at the companies where there is this feeling of a leveled playing field, and in almost all of those cases that sense of safety is created by the one super-cool person people are following. What happens is really that one form of regression and transference is replaced with another form.

**RUFUS GRISCOM**  It continues to be challenging to become a boss.

**KYLE SHANNON**  I think one of the legacies is that it created a culture and a work environment where you don't punch in and punch out. You don't say, "I'm taking my coffee break now," and hear, "Be back in fifteen minutes square or we'll give you a demerit." It can't be that way, because someone will work for twenty minutes so intensely that they need to take an hour break—or someone else will be just kind of dicking around for eight hours, take a little break, and come back and put in ten hours after that because they had an epiphany. And we need to create an environment that supports that. And then we need to be responsible enough to say, "If you're always staying here until two in the morning then something's wrong." We've got mechanisms now in place that measure people's billable hours, and if someone's billable-hour rate goes up above a threshold, we say, "What's going on here? Are we getting ready for a launch, or are you just burning yourself at both ends?" Because in a company of self-starters people have a hard time saying no, so they'll take on that sixth project. And they'll be here until four in the morning for two weeks in a row. And that's not healthy.

**CLAY SHIRKY**   About a year ago, some whiny guy came on the WWWAC [member email] list writing about how horrible it was to work in the Internet industry. A starting salary of fifty thousand dollars out of college doesn't seem so horrible to me. So I wrote back and said, "You're a big whiner."

But then other people started to post to this thread in ways that were emotionally engaged. I thought, Well, maybe I'm wrong—maybe there's something here.

Anytime I hear anyone talk about "the working conditions" or "the egos" or "Oh, the sex and drugs," I think, I used to work in the theater. This is nothing. This is nothing compared to other industries.

But what was resonating with these people who were posting was the lack of control they felt over their lives. And I realized: Right, sure, if something's going on I can call Kyle up. But Kyle will return my calls because he and I worked together when there were forty-five people in the industry. If you joined the industry in '98, when it was a normal thing to leave college and get a job in the dot-com industry, there might be a thousand employees in your company, and the CEO wasn't necessarily going to return your calls. It might even be considered inappropriate for you to bypass two layers of management to talk directly to the CEO.

The org chart at the original Agency.com was: Kyle and Chan sat here [points to middle], and the rest of us were out in the pit, and that was it. There was no "so-and-so reports to so-and-so," no time sheets.

Every twelve-person business is unique, and every hundred-person business is the same. Because they all get the same management problems, and they all gravitate to the same solutions.

Following that string was really kind of painful, because I realized that the Alley I worked in and helped create was no longer the same place.

**BEN SILVERMAN** | EDITOR, DOTCOM SCOOP | Somebody put up a flyer in Williamsburg, Brooklyn, one day: "Interested in surfing the Web for a living? Want salary, stock options, health care?"

I'm like, Yeah!

So I sent my résumé to this company called bla-bla.com and took the day off from my temp job. I was lying in bed sleeping, and at about 4:30 the phone rang. This guy was like, "I read your résumé. Can you come over for an interview now?"

So I'm like, "Yeah, cool, let me put on a shirt."

He's like, "No, no, no, no, no. It's very casual."

I'm like, "No, no, let me actually put on a shirt."

I went to their office by Port Authority. I walked into this big warehouse space, and it was all concrete floors, exposed pipes, wires everywhere, a bunch of computers, and metal chairs. It was a converted sweatshop, and the chairs they were sitting on were the old sweatshop chairs.

**ANNA WHEATLEY** The idea that everybody's equal—well, you know what? Reality is going to set in, and they're going to find out that somebody hires people and somebody fires people. And you know what? You're not equal to your boss. And there *are* bosses. And there are bosses for a *reason*. It doesn't mean we have to go back to the pyramid structure, but that's the challenge.

**RUFUS GRISCOM** I'd spent enough years in book publishing to know that, basically, working in the real world sucks. It can be really miserable and soul-sucking. It's worth doing an extraordinary amount of work, and not being a total purist, to create something that's really going to affect people's lives, and create cool jobs for decades to come. I always had a relatively pragmatic view—maybe more so than some of my peers—and when you look around, the people with pragmatic views got further than ones with less pragmatic views.

Starting my company with friends was fantastic, but the result was definitely not to hire any more friends. There were a half dozen different friends of mine that wanted us to hire them, and we didn't. That was very hard, and it compromised some friendships, but we realized that it's a very dangerous thing to hire friends.

Like many companies, we went through—and are still going through—a process of evolving from being a collection of friends making decisions somewhat collectively to a more hierarchical—heaven forbid—corporate environment. You start relearning the lessons.

The whole idea of having management structure and full-time managers seemed totally ridiculous, like a hysterically inefficient use of human resources. Why would you have this whole management structure? You're not going to manage people. But it turns out that it's totally necessary, and that there was some reinventing the wheel that went on in many Internet companies.

**BEN SILVERMAN**  Most of the management at Bla-Bla.com—the VP of Bus/Dev, the VP of Marketing, the VP of Ad Sales, and the CFO—had all been brought over from CDNow. Once people realized what CDNow was about and where they came from, we knew we were in trouble. Why bring over a team of people from a failing e-commerce company that has totally fucked up every opportunity to capitalize on its market position?

There was a real culture clash between the typical young dot-commers with no experience and the senior management who were absolute idiots—who didn't understand what the business model was, who our consumer market was, who our business-to-business market was. Within a month of all the new VPs being on the job, we began a campaign of terror with our CEO, Hagai Yardeny. Every day, we would say, "Get rid of these fucking people, they're idiots." It was unfair to him, forcing him to deal with us low-level employees on this. At first he was very nice about it; then it got to the point where he was just like, "Fuck off." A lot of people would have just fired us.

But it got to the point where we eventually exposed what idiots they were. Some of the VPs did it themselves—they proved they were idiots. They pulled some dumb shit and thought they had more power than they actually did, and they got canned.

*How did you expose them?*

**BEN SILVERMAN** It wasn't too hard to expose—every time they did something stupid we let everyone know about it. I'm trying to remember some of the shady shit we did. I would go into their hard drives, and read their email, and read their documents.

There was internal sabotage. We did everything we could to make these people look bad, because we felt it would take too much time for them to expose themselves. We would have meetings and be absolutely nasty to their faces, disregard anything they said. They would ask us to help them work on stuff, and we would ignore them. And just constant bitching to the senior people about them.

**JON EFFRON** When I first started working at eGroups, one of the guys in the interview was filling me in on start-ups, and how start-ups grow, and how they're valued by larger companies. He pointed to GeoCities, who were trading at $100. They were vested fully, and four years vested at $100—you do the math. That whole trick, "You do the math." For all I knew it was trading at $60, but it sounded so incredibly sexy to me that I kept thinking about my 7,500 shares. If I'm getting all those at $100 apiece, that's $750,000. And, while I've never used Geo-Cities, we're a lot cooler, and it could be as high as $120, and what if they give me more stock later on? In my mind, I had already purchased a condo on Fifth Avenue.

If you look at it like a chart, the condo on Fifth Avenue is where I started my first day at the job, and from there it just went *down down down down down* . . . until I'm like, "Shit—I've got to pay my rent. What happened to my condo on Fifth Avenue?"

**CELLA IRVINE** Let's talk about diversity and the complete lack of it in this industry. If you said to me, in what ways do I feel we haven't done a good job? I think we've done a really good job of making money and setting a foundation down for what's going to be a phenomenal industry; we have a great talent pool, et cetera. Partly because the industry has created personal wealth rather than company wealth, we have done a terrible job of reaching out to [underprivileged] people who live two miles away from here. There is a willingness to believe the free market will just make it possible for everyone to have a computer, and for everyone to have equal access, and it's just not true.

**SCOTT KURNIT**   Someone came and joined About.com, and we handed her her laptop, and our guy in desktop support said, "I'll get you a case this afternoon," and she said, "What do I need a case for?" And I just couldn't believe it.

It's not that we mean to kill our people. We don't. We're not a sweat-shop—but we mean to actually free you up while having you work all the time.

At one time, when I covered payroll once or twice for sixty-one people, we put everyone on half pay, and a bunch of us went back on no pay for three months. The strength of our company is that it never leaked that we were on half pay for three months—whereas in most companies that would be the first thing to show up all over the place. And I think that's a testament to what we were building, and the belief people had who worked here. Of the sixty-one people only one left, and that individual needed to for financial reasons, and it didn't leak.

**ANDREW RASIEJ**   And in the end I think—and this is more of a personal comment for myself—that I'm a producer. I don't feel like I'm alive unless I'm producing. There's just so much opportunity, I feel like I'm on the Mason-Dixon line, and they just fired the cannon—there's this incredible real estate out there, why wouldn't I want to just gobble up as much as I possibly can? I don't mean to sound so greedy, it's just that the opportunity's so great, it's actually this unique moment in time. I literally jump out of bed in the morning. I can't wait. I'm looking for technology to find a way to cut my sleep time safely from six hours to two.

**SCOTT KURNIT**   I learned a lot of lessons at the various places I worked, and one of the lessons I learned at MCI was, this was a company that worked all the time. This was a company that handed out laptops to everybody—this was before About. This was a company that checked its email on weekends.

While AT&T went to sleep Friday night and woke up again on Monday morning, MCI was busy cooking up ways to compete with them over the weekend. Raising an issue on Friday night at ten o'clock, you had it completely resolved on Sunday night at ten, before AT&T even knew what the issue was. I mean, it was a lesson for me: conventional corporate companies *can* run really hard. A forty-thousand-person company, and it ran twice as fast as AT&T. This was six years ago.

That means that *we* can run twice as fast as other companies. As long as we have the people here who get a kick out of doing that. So we do our best to get people here who do. Run faster, run harder, be smarter, be better, get compensated for it, get rewarded for it—you'll have a better company. Figure out how not to burn people out in the process. And that is possible.

Internet companies, as we get larger, are a lot different at five hundred people than we were at fifty in terms of how quickly we can move. I don't think we've slowed down. You start to think of other ways to work, to make sure you don't. I mean, we do orientation today, when with fifty people you didn't think to do an orientation. I spend an hour and a half talking to new employees—about the fact that I expect them to answer email on Saturday and Sunday, for example. And I probably should have told you that before you got here, I say, but if you don't like it maybe you shouldn't be here. And I'd rather you know it now than in six months, after you've invested time and we've invested time.

The reason we give you a laptop is that if you take Monday off that's okay, because you can work also. You can do it from a beach house, and you can be productive on your own schedule. We have nomenclature around here: WAH, WOV, and ROV. WAH is "work at home," and you just see it in the subject line. And "work at home" around here means you're really working. It means that you're by your computer or at your phone if you want to get contacted. WOV is "working on vacation": I'm on vacation but I'm such a freak that of course I'm going to check my email a couple of times a day, and you can reach me on my cell phone. ROV means "really on vacation." ROV is important, and I discovered why this past Fourth of July weekend, because it was the first time in four years that I had been unavailable from email for more than six hours.

We do want people to take time off, because if you don't you're going to burn out.

**JOSH HARRIS** Kurnit's an interesting animal. I actually like old Scott. He's kind of the jerk that you love. He's a businessman, he did it, he's got a nice point of view, and he'll do what he has to do to make his company go forward.

I was going to give you my "not sure I'd want to be in a room with him for a long length of time" line but actually he would be okay. He passes the small-room test, barely.

**CLAY SHIRKY** Employees feel betrayed because the openness and flexibility of the culture that characterized the Alley in the early days simply went away under the pressure of running for-profit businesses with fifty employees.

And it will never come back, because people who quit out of that environment to start their own company found that they were now competing with their former businesses—who knew perfectly well that the upstarts were their competition.

*What is it like to fire people?*

**RUFUS GRISCOM** It's fantastic. We've fired several people and it really changed the atmosphere in the office. You realize that people need to be told that they're not doing a good job when they're not doing a good job, because it's the only way they'll believe that they are doing a good job. People like to feel someone's noticing what they do.

It's hard for people not to be managed. It's exhausting and depressing. It's like not being paid attention to. And if people beside you aren't working hard and no one seems to notice, it just destroys office morale. So firing people who are in those positions is often a therapeutic process. It's good for the people who get fired, too, because they realize they have to change the way they're working. It's more fun to work in an office where people are pumped and driven and getting things done.

People came up to me afterward and said, "Rufus, I don't feel a lot of job security." Exactly! If you want job security, go work for the government. We're not trying to create a place where there's job security.

# LIFE INSIDE THE
# BELLY OF THE BEAST

*New York City was the perfect stage for the remarkable events that occurred in Silicon Alley. One of the reasons so much was able to happen in the Alley in such a short period of time was the convergence of different industries: the proximity of ad agencies, publishing, the press, and Wall Street all helped to speed the growth.*

*But there were other factors—the people, and the fact that New York wasn't traditionally a digital town. In the tech world, New York has often been criticized for being behind the curve; without one of the top three tech schools in the country close by, many people, including investors, were skeptical about investing money here. The experience of growing an industry from nothing in the hostile climate of New York City profoundly shaped the experience of the entrepreneurs.*

*The rise to power of the entrepreneurs was also a classic New York story— young people with big dreams climbing their way to the top.*

**MARC SINGER**  It's cool starting a company in New York. You deal with things here that you don't have to deal with anywhere else. If you can get an office space, it's like you *deserve* to go public: How did you *do* that?

We had this guy who was the super of our building, in our first real office space. You remember *One Day at a Time*—Dwayne Schneider, the super? This was like the evil Dwayne Schneider.

There'd be a leak in the ceiling, and we'd be like, "Terry, can you call the landlord to fix the water that's dripping on our computers?"

He'd be like, "Hey, you got fifty dollars?"

I'd be like, "Terry, we pay rent, there's water—"

"Can you buy me a six pack?"

People also tried to break into our office once; I was at the office at three in the morning with the NYPD finding out all about how people break in. Life in New York is hard.

**CLAY SHIRKY** That first place we were in was on the second floor of 1234 Broadway, and we went in, and we tried to pull bandwidth; this was when ISDN was unimaginably expensive, so we thought, We'll just pull a bunch of modem lines. And they said, "There's no more twisted-pair [insulated copper cables] in the basement."

"It's a huge building, how can you be out of twisted-pair?"

"Oh, the executive recruiting firm down the hall is using sixty phone lines."

"Sixty phone lines?"

Well, it turned out it was a phone sex operation, and also a brothel—now defunct. They called it "executive recruiting," and there was a sign outside that said INTERVIEWS HELD DAILY, which I think was a way of explaining the guys in suits and ties who were coming in and out at lunchtime. I think interviews were not the only thing held daily down the hall.

**JASON CHERVOKAS** First and foremost, New York is obviously a media town, not a technology town. And I think New York proved to the world that this is a media business, not a technology business. This is about connecting people to one another in new ways and creating new kinds of businesses and methods of communicating around different kinds of messaging and media environments. And the technology was really incidental to that. That's sort of New York's gift to the world.

I really believe that that message wasn't coming from anyplace else in this Internet economy. Clearly New York was then, is today, will always be less of a technology place than Boston or San Francisco. It's the nature of the culture of those places: central institutions—whether it's Stanford or MIT—attract a certain kind of talent that wants to stay there. New York attracts a different kind of talent, and it has for years. Creative talent—writers, musicians, actors. There are different human resources here, and there are also different industries here to piggyback on—financial services, advertising, publishing.

**ANDREW RASIEJ** New York is a publishing and marketing capital. Jason Calacanis was the one who sort of gave me this perspective, which I use all the time now: Early on, Mosaic could only present pictures and text, and the people who knew how to tell stories with pictures and text were already based here in New York. The artists are the people who understood how to create nuance; they were based here, so they gravitated to the Web. So application-based or content-based Web-oriented production began happening here.

**DAVID LIU** New York has actually been important in ways that were a little unexpected. When you are trying to start a business it's a very hostile environment: taxes are ridiculous, the costs are ridiculous. But the one thing New York has a rich resource of is people. And it's people who aren't quite educated on stock options, and that's a big difference. I have talked to some of my peers in Silicon Valley who are CEOs, and I literally don't know how they can run their businesses. There the problem is that you have neighbors who are dot-com millionaires. And keeping up with the Joneses—where you measure where you are in the world based on stock options—is so distorted in Silicon Valley.

**ALICE RODD O'ROURKE** There was and will continue to be an extraordinary breadth and depth of talent here. While New York may share world leadership in financial services with London, fashion with Paris, entertainment with L.A., there really isn't another city that has leadership in so many areas. And that is what created new media; that is what has sustained it. My phone is constantly ringing with people across the world wanting to know how this thing got started here. How they can get the things we're talking about today started in their own province or district or city. That is what has always made New York strong, and hopefully it will continue to do so and keep this industry strong here. Are there threats? There are, the same ones that went against biotechnology—the cost of doing business, the cost of people. And the lifestyle here is really not for everyone. It's rough. There are days when you say, "Why am I here?"

**KEVIN O'CONNOR**   We had people in New York, Silicon Valley, and Atlanta, and the question was where we should form the headquarters. The obvious choice was Silicon Valley, which was the home of the Internet, but our belief was that Silicon Valley was the horizontal technology—we were really a vertical technology. We were trying to change the way advertising was done. Our customers were publishers and advertisers, and New York has an overwhelming concentration of that—Silicon Valley doesn't really have any.

At first I was totally against New York. My first experience here took place in the late '70s, early '80s, and it was all very negative. I pictured a city that hated companies and hated start-ups and hated work. But when I came here I saw that the city had completely changed. Guiliani had just turned it around, and the city was incredibly vibrant.

**SETH GOLDSTEIN**   There are just not enough technology people in New York. You don't have a talent pool like you have in Silicon Valley. You have a lot of good marketing people who pretend they understand technology, but when it comes down to it the emperor has no clothes.

**JASON CHERVOKAS**   Actually, New York did always have a big chunk of programmers who were network database application guys, and they all worked down on Wall Street at investment banks and at stock brokerages. This was the absolute backwater of the programming world—the least sexy, shittiest kinds of jobs you could have. They paid really well, but it was just not the glamour job. Then boom! Along comes the Internet, and suddenly this is the appropriate skill.

New York actually had a lot of talent. It's just that in the early days it was very hard for the start-up companies here to recruit those people who were making several hundred thousand dollars a year on Wall Street. Companies hadn't gone public, so there wasn't the promise of options. *Yeah, you're making two hundred thousand a year, we'll give you thirty, but come work for this great company.* So it was very hard for them to recruit the talent. But there was some talent here that was capable, that had the requisite skills.

**DAVID LIU**   Having writers, editors, sales people, IT people who are truly impassioned about creating an organization as a whole, as opposed to getting their next bigger paycheck—or asking when their stock is going to shoot up—really changes your business. New York is more of a cultural center. We approach things from a multicultural, multiclass standpoint, and that has allowed us to grab a larger audience and become more relevant to more people than the people who are coming from the straight and narrow.

**TODD KRIZELMAN**   We moved to New York because we thought that, if we were truly going to be in media and advertising, this was really where most of those dollars were getting spent. Over fifty percent was certainly being spent here, and we would have an edge. We figured most of our competitors were going to have satellite offices here. We said that it would be important for us to be here, that we would be closer to the clients. That was the main reason. Then we discovered that there was a huge population of good salespeople, good marketers—people from the product side, designers, also our connection to media. So TV executives like Dean Daniels—he was working in live television, which had the same immediate nature as the Internet— were a good fit within the company.

And it certainly was a benefit that we were here at the beginning of the Internet industry in New York. We worked with DoubleClick, for instance, in the beginning of '96, when we were first just flying back and forth between Ithaca and New York, when we were just moving here. Being able to have some of those relationships very early on was helpful.

**CELLA IRVINE**   New York has always been a hotbed for talent. There is nothing like the brains, creativity, and ambition of people who

choose willfully to reside in New York City and put up with all the shit
you have to put up with to live here. But it's not technical talent, which
is a huge problem, because we just don't have a technical university
that is providing a steady flow of highly trained technology graduates.

**JOSH HARRIS**  New York City as a political entity has been a negative influence. Cost me time, money, energy.

For two years Giuliani did good things—he got this place straight-
ened out. And then he went the other way. In the biggest prosperity
boom of all time, the city couldn't celebrate or relax and enjoy its prosperity. It's been kept down.

**SCOTT KURNIT**  New York City and New York State gave us five
million dollars in tax breaks to stay in New York proper. Not to move to
New Jersey, not to move to Connecticut, which are certainly possibilities. So I give tremendous credit to the Giuliani administration, to
Pataki for doing the right thing with us. I think that was the right thing.

**NICHOLAS BUTTERWORTH**  We had this beautiful view of the
river, but it was no good because the glare from the river was terrible . . . the glare would kill working on the screen, so we had to make
big blinds to shut it all out. And then people would go up on the roof to
smoke pot, which they called "doing research." They'd come up and
say, "Hey, wanna do some research?" or "Got some research?" And
when you left the building you would have to bang on the door for a
while before you went out so the rats could all run away.

**STEPHAN PATERNOT**  That cultural aspect of New York was
very much what Todd and I both wanted. We absolutely did not want
to be in any one-track town, where all you talk about . . . you look left
there's a pocket protector, you look right there's a pocket pro . . . and
everyone's got thick glasses saying, [dorkily] "Hi! I'm Excite. I'm
Yahoo!" Having a town where everyone's Internet has its benefits;
everyone's in the same space. The downside is, if you're gonna make a
life out of it, you need diversity. You want bankers there—and, by the
way, that became very handy for us; you want people who are in the

arts, you want marketing people, you want people who have traveled from around the world. I mean, that had an influence throughout the site—we wanted to make absolutely sure that ours was a site where people could chat in whatever language. It was great for our personal lives, and it was great for our business.

**ALAN MECKLER** I think, if anything, New York was part of the hype, and in essence caused a lot of the ideas to be born and spun. Because there's so much money in the media, so much money for new ideas in and around New York. New York and Silicon Alley have been part of the hype—and part of the problem, which we've been seeing now for several months in the financial community, and which will continue, and for the good. Obviously, profit is now starting to be more important than growth.

**KEVIN RYAN** It's easy to forget now that people didn't take New York seriously as a high-tech location. In the beginning, we actually had to fight against that when we were pitching West Coast companies and competing against West Coast technology companies. Clients would say, "Well, you're from New York; you must not understand the technology."

And so I think it's been helpful in subtle ways for everyone to have a very successful multibillion-dollar company that manages data all over the world headquartered in New York City. There's great talent here, but I think it was underrepresented and underappreciated. We've played a big role in changing that. Look at the law firm that is most successful in doing IPOs, or even look at a successful investment bank—it's the ones we chose early on. It was Brobeck and Goldman Sachs. And at the time they weren't actually number one in the Internet area, after that a lot of companies started using them both. That's just a very concrete small example of how precedent was set. We see it all the time, with people calling us for real estate where we were, and coming over here now to the West Side since we've been here. We have people calling us all the time saying, "How did you handle this? What did you do here?"

Our role has been probably a little bit bigger because we're not a consumer company, so our clients are other Internet companies. As a result, if you look at all the leading companies here, whether it's About.com or iVillage or TheStreet—I mean, they're all DoubleClick clients. And so we have relationships with them. And we work with them. Most of the big content—TheGlobe, HotJobs—we work with just about all of them. We have good relationships, so we do talk to them and trade ideas on how to get the best office space and things like that.

It's been incredibly fulfilling to watch the community develop. I remember when there were twenty people in the whole city involved with it, and now there's gotta be two hundred thousand or something. Over time what you'll start to see is more and more second-generation companies. To date we've had very little turnover at DoubleClick, so I guess we haven't—fortunately or unfortunately, depending on what side you're on—seeded a lot of other companies yet. But over time that will happen.

**THERESA DUNCAN** Silicon Alley was able to be really high-profile because of the proximity to the media. And new media actually got a lot of talent from old media. There's an enormous talent pool here.

**ALICE RODD O'ROURKE** If not having a Stanford or MIT is an important difference between New York and the West Coast, another is, of course, thirty or forty years. I'm sure we'll be able to do amazing things in thirty or forty years here, but having said that, one of the biggest needs in this industry are workers—people who understand technology at every level. From people with high school degrees to people with advanced degrees in computer science and computer engineering. The lack of a Stanford or an MIT affects us not only in terms of the number of people, but I think in the vibrancy, the robustness, of the innovation investment engine. For instance, I have given a couple people in the industry unsolicited career advice—Robert Levitan among them. Robert is one of the few second-generation new media entrepreneurs that we have—on the initial team of iVillage, and now as a cofounder of Flooz. I said, "Robert, you make whatever amount of money or what-ever impact you need to make; then I want you to teach at Columbia or NYU. You have your choice, a semester or two. And then you have to start your third business, taking six to eight of your students with you, who will then do the same thing you've done."

The solution isn't only second- and third-generation new media people going back and becoming professors; the solution will also lie in someone or some organization stepping forward and taking a leadership role in bringing together industry and academia. It lies in industry—the Merrill Lynches and the Goldman Sachs and the Met Lifes that rely on these technologies, as well as the dot-coms—bringing them all together and taking a leadership role. Because that is really what it will take. And it'll be complicated. It'll be very difficult.

**KYLE SHANNON** New York doesn't let you stop. New York is relentless and ugly and awful. And it drives you to be creative, and it drives you to be successful, and it drives you to not fuck around. Or, if you are going to fuck around, you have to do it in a world-class way. The drive is a piece of it. This is the center of culture, the center of business, the center of world communication and trade. And a website is kind of an epicenter of some of that stuff. Some of it's culture and content, and some of it's very businesslike, and some of it's international in scope. It feels like a city that's got a lot of skill sets that can support this kind of business. I couldn't have started this anywhere else.

**JASON CHERVOKAS** What's happened is, companies have grown, and the price of New York City real estate has always marched in lockstep with the stock market. It throws off so much cash into the local economy; people go home with huge bonuses; it drives the tax receipts; it drives capital spending among corporations. So if the precondition for the creation of Silicon Alley was the economic disaster of the late '80s and early '90s, the challenge for the Alley was the economic boom of the late '90s, where real estate went from fifteen to forty dollars a square foot, right at the point where companies were outgrowing their early-stage success, and needed both to recruit and to move.

**STEVEN JOHNSON** On the East Coast, the idea of computers being of interest culturally was just a nonissue. *Wired* had clearly understood it, and the San Francisco people understood it. Obviously there were tons of people who didn't really know how to use their computer, but even the expert users weren't using it other than to make sure their spell checker was checking better. The idea that there was this art form, this expressive form, was totally new. That's one of the reasons *Wired* was such a revelation, because there had been this pocket of people thinking about it over here, and then suddenly this missive comes from the West Coast. Whatever else you think about their politics, they just *get* something here.

**ALAN MECKLER**  For New York City, the most exciting thing has clearly been the availability of money and new ideas. Obviously the Internet industry is responsible to a great extent for the jump in real estate prices, for the salvaging of buildings that were derelict, for reviving downtown areas that had heretofore not worked. Look at 55 Broad Street [the former Drexel building] as a perfect example—not a derelict building, but the Rudin family has done quite a job there. And there are other buildings like that that have been kicked into high gear and modernized because of the Internet.

I think the money that's come into New York City, and the development in the city as a job haven for the real estate industry, for the media industry, for advertising—it's been quite incredible what the Internet has done. By the same token, there could be problems down the road, because there's probably overbuilding now. While there's a lot of money available for new ideas, there are a lot of big spaces that may be empty in the next few years, too.

**ESTHER DYSON**  New Yorkers in general tend to be more diverse and have a broader world view. Silicon Valley is so much more narrow. Even if you come from India or you come from somewhere else, you start to live in that world where everything is high-tech, and when you go to lunch people are always talking about routers or websites or search engines at the table next to you. Here, you go to lunch and maybe they're talking about finance; maybe they're talking about fashion; maybe they're even talking about a novel. And I think that's good. It's why I like it here.

The other thing is: I got here at five this morning, coming back from London at a very odd time, but New York generally is open twenty-four hours a day. People may be working in their offices twenty-four hours a day in Silicon Valley, but the community is not: you can't find anything unless you get into your car. It may be New York and it may be rarefied, but it's closer to the experience of the rest of the world than Silicon Valley is.

Silicon Alley has been shaped by the proximity to the financial community and the rest of the world. To the extent that the Internet is now global, we're less American here in New York; we're more cosmopolitan, whatever that means. I still think Americans who come to Europe don't understand it well enough, but they have a better chance coming from New York than from California. And California's very money-driven. Wall Street is, too, but somehow there is a sense that there are other things to life here in New York.

**JOHN YOUNG**   New York is just a society of ideas. That's what we get off on. Regardless of whether it's print or Internet or wireless, which it will be tomorrow, we're just a society in New York that relishes ideas, that survives by ideas. If you don't have an idea tomorrow you're dead. That's the pressure of New York. You have to constantly be at your best, outdoing yourself. You have to be restless with the status quo, and throw away what was yesterday, and constantly reinvent yourself. Once you stop doing that you're dead. You might as well go off to the Berkshires or something.

**MARC SINGER**   There are a lot of huge advantages to being in New York, but it's not easy. You can be sitting in your office some day, and Jason Calacanis calls you up and says, "I've got the *Good Morning America* film crew; we're just walking around Silicon Alley and wanted to pop by, see what's going on."

I'm like, "Okay, that's fine. We'll just cancel this meeting. Bring them by." So there's that New York surrealism: anything could happen here, and it does.

**CRAIG KANARICK**   I used to joke that New York was like the Internet. Totally global, but everyone speaks English; you need a lot of money to really get around well, but even if you don't have any you can still find a lot of interesting stuff. There's lots of hidden surprises. The infrastructure is mediocre; it pretends to be organized when it's really not.

**JASON CHERVOKAS**   New York definitely has infrastructure issues, which have an impact on all the older northeastern cities—they were built a hundred and fifty years ago, and it's a pain in the ass to have to cool them, to run electricity wire. New York City was out of fiber, the capacity for fiber, south of Fifty-seventh Street as of a couple of months ago. And they actually found some unused ConEd conduit they could run fiber through. But it was a literal bandwidth problem. That's a problem that the entire old industrial Northeast has, that I think younger Seattle certainly doesn't have, for example. The cost-of-business challenge remains here for companies at that growth stage.

**ALAN MECKLER**   One of the key things about this is the geography itself. New York City is a much smaller space than what you have out in San Francisco and in the Valley. Then, too, obviously VC investors of Silicon Alley are different than on the West Coast. The West Coast could just throw an idea at them and they'd go for it a lot quicker. In Silicon Alley, the people are more cerebral in their investments. The people in California are clearly more laid-back than the people in New York, but I don't really see much important difference in meeting with people.

**ALICE RODD O'ROURKE**   This has been New York's first big industry since television. There have been other contenders for that spot—most notably biotechnology, which had as much going for it here in New York as new media did. You would think that new media is a layup, just natural, but it really just is not. I think that it's instructive to stop to consider biotechnology as an industry. New York has all these major hospitals and research centers, a population to test on, capitalists who were investing in the area, et cetera. Biotech had a chance to make it here in the 1980s, and didn't. And why? There are a lot of reasons. One, the cost of doing business in New York, with taxes, real estate, the cost of people. The cost of attracting top researchers here from other places. When VCs started investing in biotechnology, they said, "Not with my money, not with those taxes, not with that cost. Why don't we just move on over to Virginia or to one of the other middle Atlantic states that were interested in giving tax holidays for ten years?" So you had New York VCs saying they wouldn't pay this money, and you had a very aggressive economic development policy from other states saying, "Come with us and we'll make it worth your while in lifestyle, tax forgiveness, and other things."

**MARC SINGER**   I'm definitely tired. It's been incredibly intense and grating.

The energy in New York, which is layered on top of the Internet, layered on top of new media—it's a really intense experience. These first five years, you're in a cab, going to a meeting, your cell phone's ringing, you're writing something in your Palm Pilot, another guy you founded the company with is sitting next to you on his cell phone, the cabbie's on a cell phone. The intensity of living life Internet speed in New York City—you definitely could hurt yourself. There are times, on airplanes and stuff, at six o'clock in the morning when you're running into people just thinking, Someone's going to have to crank this down. Someone should crash NASDAQ. Let's do it in April! We gotta slow this thing down, because someone's gonna hurt themselves.

A couple of people told us to move out West when we started and I was always glad we didn't. New York City is a huge, huge character, the city should have been one of the Silicon Alley 100, because it made the whole experience. I go into Johnny's on Twenty-fifth street, across from my office, it's this narrow little sandwich lunch place run by Johnny and his father Larry, and can bump into three different people running three different companies.

**JASON CHERVOKAS**   The sense of specialization crept in slowly but surely over two or three years, to the point where we realized, Maybe New York isn't ever going to be the strongest on the technology side—that another place will be—because it just won't have the same labor pool. But that's okay. That doesn't mean it can't work.

**ANDREW RASIEJ**   Two years ago I was the first David Rockefeller Fellow to come from the new media industry. This is a group of ten people every year who get selected by the New York City Partnership to go through what they call a civic orientation program. You meet the chancellor, you meet the mayor, and they sort of teach you how the city government works. Whenever I was in those meetings and meeting these bureaucrats and politicians and other business leaders, it was amazing to me how slow-moving, how disconnected the traditional foundations of New York City business and communication were, and how oblivious to the opportunity associated with the World Wide Web and the energy level and the entrepreneurial spirit that had been the religion of potential technology. Now that's less and less the case, and companies like Time Warner are morphing as we speak and getting the religion. But it's still a very slow process.

**BRIAN HOREY**  The Internet had a profound change on New York, and it will be a long-term change for the better. What it really did was open up people's eyes to the whole idea of entrepreneurial wealth creation, something that really had gotten ingrained into the culture of the West Coast and was practically in the water that people drank when they got up in the morning, but that people here really didn't appreciate or understand. Historically this was a town dominated by Wall Street and Fortune 500 companies headquartered here, and people were compensated by salary and bonus. Stock options—particularly in early-stage companies with the potential to grow very rapidly—were something people tended to discount or looked at as very risky. I remember trying to recruit executives out of these companies in '95 or '96, and it was very hard to convince them of the opportunity to make a lot of money in equity. But if you flip over to '99, the first question out of people's mouths was, How many options do I get, and how much of the company is that?

**MARK STAHLMAN**  This is Babylon. This is the capital of the Beast. And if you want to do something about the Beast, this is where you have to do it. To the extent that people want to slay dragons, New York is the obvious battlefield. And what has been marvelously complex about the way this has all developed is that there have to be a lot of people who want to become dragon's pets in the process. And there's been plenty of that, and will continue to be plenty of that.

# PAPER
# MILLIONAIRES

*F*ive years after starting Razorfish, Craig Kanarick, thirty-three, was worth more than $200 million. He bought a '65 Corvette Stingray on impulse off the street, and was photographed by the New York Times hopping over the door of the blue convertible sporting a flashy red suit, open-collar shirt with no tie, and peroxide-white hair. On a sweltering summer day, Kanarick commandeered an ice cream truck and treated all of his employees.

Agency.com IPO'd eight months after Razorfish, raising $77 million to fund a dozen acquisitions by Kyle Shannon and Chan Suh over the next six months. With an IPO that raised $153 million and gave the company a valuation of $2 billion, Suh was named by Fortune magazine as one of the forty wealthiest Americans under forty.

Two years after meeting with Flatiron Partners, StarMedia was a public company worth over $1 billion. Jack Hidary's EarthWeb was worth $400 million. DoubleClick, now the biggest online advertising network in the world, was worth $8 billion, making it the wealthiest company in the Alley.

By late 1999, more than twenty-five Alley companies had gone public, rais-

*ing billions from the public markets, and turning their founders and top employees into millionaires many times over.*

*But the vast majority of the new millionaires were rich only on paper. Unless they sold their personal stock—a move that would suggest a lack of faith in the future of their company—the founders could only wear their wealth as a calling card, not use it. They lived fast lives on little sleep, worked 'round the clock for years, and saw their tinsel fortunes multiply until the end of the century.*

**OMAR WASOW**   I've always had a thing for business and for hustling a little. I grew up in Manhattan, on Bleecker between Mercer and LaGuardia, and when I was in elementary school I went to Washington Square Park and sold lemonade, and I actually got busted by the police for selling lemonade without a license.

In high school I threw parties where I charged people, and in college I sold T-shirts and made hats. I've always enjoyed business and the process of trying to take an idea and make it real. I also grew up in a family of teachers, so for me there was no greater rebellion than going into the world of mercantilism and actually getting my hands dirty selling stuff.

**KYLE SHANNON**   I knew I wanted to come to New York, so I moved into East Harlem on 114th Street and First Avenue. It's very hard to be here and not be driven to do something, so pretty quickly I started and ran a theater company, I designed a chess set, I designed a piece of software. I wrote seven screenplays in two years, and then I started a management company, which my wife ran, to help us get those screenplays read by producers. I've always been hustling up something.

When you do a show it's just a hustle. You gather the people, you somehow gather the money, you make the time, you put it together, and you make a show.

There's something about the city that rewards hungry people. It rewards people who are willing to be entrepreneurial, take the ball into their own hands.

I had created five or six businesses or ideas or companies while I was still acting, and nothing took. It was all crap. And then I started

UrbanDesires, which led me to start the WWWAC, which led me to start Agency.com. And I thought, here are three things that are possible, and one of these might make it. Of course, Agency is the one that really took off. UrbanDesires and WWWAC were pretty powerful voices in the early days, though, and they're still around today.

**CLAY SHIRKY**  Seth Goldstein, Kyle Shannon, a bunch of other people from the early companies, and I had all come from showbiz. Partly because we had nothing better to do than sit around and wait for the agent to call, partly because people who gravitate toward that industry typically have a what-the-hell attitude. They're used to not being scared by things they don't understand. That's what rehearsal is for: You're used to this period during which you don't understand something, and then you get together and talk about it really intensely, and at the end of it you do understand it.

But it also meant that we tended to run things on a project-by-project basis, and that kind of emotional intensity within the group was unsupportable as a way to run a long-term business.

You can run a rehearsal where everybody screams at each other, slams doors, then kisses and makes up and says it was all the most fabulous experience of their lives, because after the opening-night party they're only going to have to see each other two hours a day. But if you live in that kind of emotional hothouse all the time it really gets tough. And that was something we had trouble with at SiteSpecific.

**SETH GOLDSTEIN**  Part of starting a company is hype, because there's nothing there, and you're always trying to pretend there's more there than there really is.

**FERNANDO ESPUELAS**  StarMedia launched in 1996, and we were the first service in Latin America that had chat through the Web. But in the very early days there were very few people on the Web and even fewer who had found us, so I spent twelve, fifteen hours a day in our chat rooms pretending to be different people to create a critical mass.

**SCOTT KURNIT**   I started on no pay for the first year and a half. I had the support of my wonderful wife in that, and obviously we wouldn't have existed if she wasn't game for it.

I went on payroll for all of two or three months, and then went back on no pay. When I told my wife, she said, "Are you sure this is a good idea?"

Your wife is really your partner, and it's a lot more grueling to do this alone.

I had come out of traditional corporate jobs where I got a big pay-check and a car allowance, and it was very cushy. I was a division president of MCI, so I had jet rights. I'd just call up and I had a jet at my disposal. And now I was trying to figure out what was cheaper than coach to fly.

**SCOTT HEIFERMAN**   This notion of the dot-com entrepreneur being a twentysomething kid is largely a myth. Ninety percent of the successful companies in the Alley—companies like Agency or DoubleClick—were started by people in their thirties or forties.

Not only was I not disadvantaged by my youth, but dare I say a little bit advantaged, because of the novelty factor. The first few people I brought on to help me run i-Traffic were really experienced people who weren't in their twenties—but very often the press didn't want to see that.

**ANDREW RASIEJ**   My father spent forty years building his career, and never made more than eighty-five thousand dollars a year in his life. I'm basically compressing his entire career trajectory into four or five years. The payoff, I think, might be being financially independent enough or experienced enough to be able to take advantage of other things in life that our parents never were. That's the promise of technology.

**JOSH HARRIS**   We'll see how fast I can blow all my money. And I'm—well, I'm blowing money like a drunken sailor.

**ALICE RODD O'ROURKE**   New York is like a black hole. You have to be so huge and powerful to avoid being sucked in. So if you look at Todd and Steph [of TheGlobe.com], and you look at Craig, who lived large, in some way they were doing a service to the industry, because they were promoting it. Did they live too large? I know some people think they did.

**CRAIG KANARICK** I Rollerbladed to work because it was fun, and I wore bright colors because it was fun, and I dyed my hair because I was doing design experiments just like I would do on screen or on a piece of paper. Somehow it became this larger-than-life thing, and I think it would be dishonest to say I didn't want the attention.

**ANDREW RASIEJ** Jack Hidary had signed on to help out with MOUSE early, early on, and then I think he was at one of the first wiring days, and then six or eight months later EarthWeb went public. The day after he went public I got a phone call from him, and I didn't know Jack that well. He was being interviewed by the *Wall Street Journal*; it was the largest IPO in history; how did he have time to call me?

And he said, "I've always wanted to be in a position to help MOUSE financially, and haven't been able to do it. This is really important to me, and I want to tell you that I'm sending a check tomorrow for ten thousand dollars for MOUSE." And I was totally flattered.

The next day the messenger comes with the check—and the check is from his father. All his wealth was on paper, because he was fully locked up. Here was this guy who was being touted as a billionaire, but he really couldn't even write a check for himself. This is the sort of disconnect that occurs when people try to understand what the actual practical power of the wealth really is. [Hidary denies that his father wrote the check.]

**RUFUS GRISCOM** The appearance of success and glamour in the lives of these Internet entrepreneurs was difficult on a lot of people around them, not just on friends but on people just reading newspapers.

**CRAIG KANARICK** Any time you have a get-rich story you have people who are envious.

**MARC SINGER** Certain people seemed to be very affected by the fact that so-and-so had a lot of money. And it's funny when a company goes public and you know the guy who's one of the founders of that company. You're like, Wow, cool, they have lots of money now. But I definitely was at parties where some people seemed to be very fixated on "Oh my God, that guy's worth forty million dollars."

**CLAY SHIRKY**  People felt betrayed by the sense that you were kind of dumb if you didn't have a million dollars. The notion was that every third person was making a million dollars.

My order-of-magnitude calculation is that one in ten thousand people who were there in '96 and '97, in the early days of the Alley, made a million dollars or more on the Internet. It's such a tiny fraction. But they were all people we knew because they were these hyperconnected Alley figures, and so it was an illusion.

The double illusion was that they didn't really *have* a million dollars, they had the right to buy a million dollars if they were to sell all their stock today—which is a very different thing, as people have learned.

**CRAIG KANARICK**  I was having fun. Every single day I was coming to work I was having a blast. The types of problems we were trying to solve for companies were fascinating to me. The ability to control my own destiny and where I was going was fascinating. The people I was meeting and interacting with were unbelievably great. The opportunities I was having to go places, to go to different cities and different countries and interact with people in Sweden and give lectures in Mexico and all these different places in South America—to be able to do all those things was fun.

**ANNA WHEATLEY**  The downside of what happened was the creation of a group of truly arrogant and self-centered people. I don't care, I'll say this any day of the week. Having worked at *Omni*, it was a great privilege to interview scientists who had won Nobel Prizes because they had worked on game theory or had discovered quarks. And these are truly events. So when you have a context for understanding really important things and you look at the self-importance of someone saying "I've created this company and my company did this on the stock market"—to me there was no comparison.

What bothered me was the lack of recognition that, yeah, this was cool, but nobody's going to remember your name. And I think Jeff Dachis was in fact the most honest of them all, and the most arrogant of them all, when he said [in the *New York Times*], "I feel completely and utterly entitled to whatever success comes our way."

**THERESA DUNCAN** The whole culture started to seem obnoxious. People really wanted to sell themselves as celebrities and show everyone how very stylish they were. It became really consumption-oriented: "Look at my car, look at my penthouse apartment." I'm not above wanting a good lifestyle, but at the same time it became just sort of repulsive to me. It was like lifestyle porn.

**MARK STAHLMAN** If you play with media on its own terms, then what you're looking for is to become rich and famous. Because you've got no argument with what's going on. And if you've got no argument with what's going on, you can't build anything new. So the effort to fit in, to be successful on the terms that media defines it, has to lead to that kind of realization—that it's impossible to be outside of it. We're all inside the simulation.

**MARISA BOWE** There are some bad aspects to having all that money floating around. The real estate situation is totally impossible. Young artists cannot afford to live here. But there were so many good effects, like people having money and work and being able to afford to live here. Being able to try things and do things in a strong economy is a really great thing, for the most part.

The side effect is that you have to be envious of someone who has a lot more than you. That's just childish. I thought it was funny and entertaining to have an asset and . . . I don't even think of it as any excess. Excess of what? If you have a lot of money, you're going to spend it.

*When Alley companies went public, the founders often became multimillion-aires. iVillage, which started as a consulting gig for AOL but grew quickly into a well-trafficked online content network targeted at women, went public in the spring of '99 and was worth more than $100 per share (though less than two years later it would fall below $1).*

*Robert Levitan, one of the three original founders of the company, had already left to start his own startup, Flooz.com, when iVillage went public on March 23. When it closed its first day of trading near $80, his stock in the company was worth tens of millions of dollars.*

**ROBERT LEVITAN | CO-FOUNDER, IVILLAGE |** I have lunch with my dad every Saturday. He's the guy who has lived the American dream. My father was born in 1909, and he's seen it all: the automobile, the radio, the TV, you name it. Forget about the telephone, the fax, and computers. His parents were immigrants, and he's a self-made man.

Two days after iVillage went public, I had lunch with him. And he said, "So, how many shares do you have?"

He's my dad, so I told him. And he looks at me and goes, "Um, you know, that's *a lot* of money."

I said, "Yeah."

And he said, "Well, what are you doing with that? Who's managing it?"

See, before a company goes public, people from Goldman Sachs (or whoever is taking you public) sit down with the executives who own a lot of stock and say, "Look, we want you to be a private client of our company." So I said, "I'm just letting the Goldman private client services group do it."

And he goes, "Well, all right!" I could see that that, like, validated his life. He felt *he* could never have been a Goldman Sachs client. It changed his whole view of me, and I think he finally figured he didn't have to worry about me.

*Did he let you buy lunch?*

**ROBERT LEVITAN**   He let me buy lunch that day. And then he let me buy lunch for a long time. And now *he's* back to buying lunch, because times are rougher on the Internet. He's buying me lunch, all right. "Save your money," he says. It's very funny.

**CECILIA PAGKALINAWAN**   Sometimes people ask me why I still seem to have energy or enthusiasm. It's because I've always tried to maintain a balanced life. I try to keep my weekends to myself. I rarely go to the evening parties, because after seven I want to keep my time to myself.

With Marc [former boyfriend Marc Scarpa, of JumpCut], basically what happened was I would work really hard to finish everything by seven P.M., and he would work until eleven P.M. So I never really saw him, even when we were living together. I wouldn't count the sleeping hours as quality time spent together, even on trips. We went to Brazil, San Francisco, Key West, and Puerto Rico, but the guy was always working.

**STEVEN JOHNSON**    Just because Stefanie and I are two different genders and we're partners, people think we're involved, though we've never been involved.

**STEFANIE SYMAN**    That's been much easier since Steven got married.

**STEVEN JOHNSON**    One time, right before we were hosting the WiredNews Bureau, Gary Wolf, who'd been a longtime writer for *Wired* and was starting up the WiredNews Bureau, was coming into town. He called me up and said, "Set up a big Silicon Alley get-together. We want to meet all these people. We're about to go public; maybe we'll acquire some people."

It didn't exactly happen.

We were standing outside a restaurant waiting for everyone to show up, and he's like, "Who's coming?"

I'm like, "Well, we've got a very good list. Stefanie's coming, and Nick Butterworth is coming, and I think my girlfriend is going to come as well."

He's like, "Great, that's good."

But I could tell something was a little awkward. After thirty seconds he said, "Aren't you and Stefanie married!?"

And he totally thought I was Mitterand or something—my wife will be there *and* my girlfriend.

It's a very liberated city.

**RUFUS GRISCOM**    People often say, "Once you cash out of this, then you can do what you really want to do."

This *is* what I really want to do. I feel unusually lucky because my labor of love has also turned out to be this sort of great financial opportunity. It's very rare that those two things coincide. You usually become an investment banker, and then go start your bed-and-breakfast. That's sort of the classic model.

I just feel absolutely grateful for my historical timing. Throughout history, people have had to do miserable things to survive and put food in their mouths, and that also rings true today. Most jobs suck. Getting out of college is very depressing. And if you can find a way to do something more interesting, then fantastic.

**MARC SINGER**   Dick Gephardt called me one day in my office a year ago.

It was a Friday afternoon after a bad week, around four thirty, and I'm thinking maybe I'll get out of the office early. The phone rang. "I'm so-and-so in Congressman Gephardt's office. He'd like to talk to you. Can I patch him through?"

I was like, "What?" I'm looking around the office like, Who's screwing with me? I don't know Dick Gephardt.

So I asked, "What will we be talking about?"

"We're doing this big rally or fund-raising or meeting or something for encryption. We know that's really important to your company."

And I'm thinking, Encryption doesn't have anything to do with our company. That's a really specific thing. We do marketing.

So he puts Gephardt through. It was a three-minute conversation about encryption. And the whole time I'm thinking, It's definitely a Silicon Alley coming-of-age moment when you have Dick Gephardt prank-calling you.

**JOSH HARRIS**   I can see in the business world people burning out. They're getting out. You burn hot and hard and you go, and you want to burn out with money and success. I burned out. I think I burned out from business.

**KYLE SHANNON**   When I go away on vacation, coming back can be pretty horrible. I remember one time we went away for two weeks— it was my first two-week vacation—to our place in Pennsylvania. Two weeks of fishing, just sitting around, doing nothing and loving it. When we were driving back, it was dusk, the sky was kind of purple, and I saw the outline of the city, and it hit me—it was like someone punched me right in the nose. I was like, "Holy shit, I'm going back into that city, where there's a company with twelve hundred people in it that I made. Can I even do this anymore?" All the self-doubt came in. It's only when you remove yourself that you can see what it really is.

**JOSH HARRIS**   For the past six months, I've had the most expensive therapist in the world. When I'm in the city I play poker. And I've been losing. I've had a terrible year. I've lost thirty, forty, fifty thousand bucks playing poker, just playing poorly. Not even bad luck, just poor poker.

**FRED WILSON**   As an entrepreneur you have to be able to create something out of nothing, and you have to be able to get other people to join you in creating something out of nothing. So you've got to be able to lead people down a path where they're not entirely sure where they're going to go, and they have to be going there based on your leadership and the fact that you can convince them that there's something good there at the end of the tunnel. The quintessential entrepreneur, that's what they're good at.

It's hard for those people to ultimately turn into really great managers and really great businesspeople. There comes a time where the company has to stop being vision and dream and *follow me*, and has to start being the blocking and tackling of bringing in money every day, managing people, making sure the expenses are under control, making the tough decisions.

Most of the entrepreneurs actually don't make that switch very well. The companies never would have existed if the entrepreneurs hadn't started them. But the entrepreneurs could never have run them.

That's much more time-honored on the West Coast, the notion that ultimately the entrepreneur has to turn it over, give the keys to someone else. Here in New York I think it's been harder for people to understand that. And I think only now are people starting to recognize that that's really the natural progression of things. Entrepreneurs don't always or shouldn't always be the ones to run their companies.

**CONNIE CONNORS** | FOUNDER, CONNORS COMMUNICATIONS | I have a lot of clients who are serial entrepreneurs, and even when they get around a couple hundred employees they're still micromanaging. They can't let go. They're editing press releases at ten o'clock at night. I just don't think you can grow a company to scale with sustainable value as a serial entrepreneur. People don't know when to get out of the way.

# THE OLD ECONOMY

*T*he Internet was disruptive in every sense, especially for old-economy companies. The sudden rise to power of a new class of digital entrepreneurs sent shock waves through the New York offices of major corporations.

The Internet affected nearly every industry, and the temptation of upper-level executives to jump ship into an early-stage start-up was buttressed by the overwhelming financial rewards to be reaped by a full-scale success. So the first to be liberated by the powers of instant wealth creation were the corporate executives who looked inside their computers and saw a mirror world, with new jobs for everybody.

Nowhere was the Internet's capacity for disruption more obvious than at the world's most powerful media companies. Bungled attempts by traditional companies to "get" the Internet resulted in hundreds of millions of dollars wasted on aborted online enterprises.

In the early 1990s, Time Warner led a vanguard of media companies in experimenting with interactive television, television that would allow you to choose videos on demand and interact with other users over a network. A collaboration with Silicon Graphics, Internet heavyweight Jim Clark's first com-

*pany, the project was shelved when it became clear that the set-top boxes would
never be affordable to the average consumer.*

*Another problem old-economy companies faced was their sudden deflation
in value relative to the pure plays (Internet-only ventures). Start-ups that had
gone public and were suddenly worth a billion dollars had instant leverage
over the traditional companies. Online media companies were valued as technology companies. Investors failed to see that even online, content suffers from
a diminishing value over time. Where software, which can be upgraded every
six months to strengthen market share, has the potential to take over an entire
market (as Windows did with the operating system market), content can only
be as strong as that day's news and features.*

*In the early days, content companies were still developing the technology
used to serve the content, and it was unclear where the value proposition
was—with the technology or with the content itself. Since the Internet was a
transactional medium, anything seemed possible—especially when e-
commerce seemed like the most promising feature of the Web.*

**CECILIA PAGKALINAWAN** In every industry, whether it's
finance or automotive or fashion, there are some people who can
claim a piece of the business away from the establishment. I think
what's unique about the infancy of Silicon Alley is that it was a huge
drove of us that forced this thing to happen and created something out
of nothing.

The big corporations only started paying attention to us when they
thought the little upstarts were going to take their market share away
from them. If it wasn't for the upstarts that tried to create these dot-
coms and pure plays and scare the hell out of the established companies, those companies probably wouldn't have gone into the Internet
for another five or ten years. The benefit, at least, of all the pure plays is
that they forced the big companies to pay attention.

**CRAIG KANARICK** Twenty years ago, in order to do your best
work and to succeed in this world, you had to conform and go work at
IBM or a big company, pay your dues, climb the corporate ladder, and
step up and step up. Over the last ten to fifteen years, the way people
have done their best work has been to become entrepreneurs and do
their own thing. I'm interested in doing my own thing sometimes. I fol-

low a lot of rules, but I also break a lot of rules. I kind of like the rules there so I can break them.

**SUSAN BERKOWITZ** | FORMER VP, THEGLOBE.COM | I walked into a law firm the other day, and everyone's in casual clothes. That came from kids like Todd and Steph [from TheGlobe] showing up to every meeting in their Globe t-shirts. Eventually people were like, "Why am I wearing a suit?" Bankers didn't want to show up at these companies looking like stiffs.

**SCOTT KURNIT**  When you realize that New York is the center of traditional media . . . while I do not believe traditional media will dominate the Internet, certainly in the near-to-medium term they may acquire their way in, just as many of the traditional media companies acquired their way into the cable business. Long before ESPN was part of Disney, and before it was part of ABC, it was an independent company. And those other companies didn't have the strength to develop it.

**OMAR WASOW**  People coming out of traditional media have the deck stacked against them, because the skills that allow you to succeed in traditional media are the skills that burn you on the Internet. The challenge is applying a content model in a technology business, and the importance of being really lean as opposed to very free with your money—as might be conventional in, say, the music industry. The three or four kids who come up in the Internet—who don't necessarily have a lot of money to start with but who really get the medium—are much more threatening than the people who are highly credentialed and raised tons of money and are these real [traditional] moguls, because those folks don't get it. As Jimmy Cliff said, "The harder they come, the harder they fall." You see that again and again.

**KEVIN O'CONNOR**  A lot of times you sit there and bang your head against the wall and say, "Why doesn't this traditional company get it? Why aren't they on the Web?" I lived through it with the PC. When I first launched ICC [an earlier venture] we were hooking PCs up to people's mainframes, and I figured that people would call us, and no one called us. I called over a hundred people, and they told me to never call again—that there was no way that a PC would ever be in their company. They would never buy PCs.

It took ten years for corporations to embrace PCs. It just takes people a long time to embrace new technology. Local area networks faced the same thing; Cisco faced the same thing. It takes people a while. Why? Because people don't understand it, don't want to learn it; people have been burned by too many fads. I think that people were really burned in New York over interactive TV. They lost billions on it, and then a couple years later there's this new thing called the Web, and all they hear is the word interactive, and they say, "Go away—I'm tired of losing money."

**DAVID LIU**  When we were trying to get people to understand the revolution we felt was happening, the example we used to sell The Knot concept was back in the '70s, when *Rolling Stone* was the de facto authority on music. I remember as a kid everybody bought that *Rolling Stone* book and read about the albums: Oh, yeah, this got five stars. You bought the magazine to get the interviews on the artist and the album reviews and all that other stuff.

Well, along came a new technology called cable television. The publishers of *Rolling Stone* probably didn't think much of it, and when confronted with the possibility of using someone else's marketing material as your content—my God, that is blasphemous! The church-and-state separation of magazines from the music industry is very important. You couldn't possibly do that.

Well, Bob Pittman walked in with MTV and realized that there was a whole generation of people who have a different relationship with television, who will consume music videos as content. The record companies will provide that for you for free. In fact, they'll *produce* that, because they see it as marketing. They've turned MTV into a global realm. It defines global culture, fashion, and other things for a whole generation of people.

I think something truly powerful is going to happen with the Internet. The integration of marketing and content is going to be blurred. Consumers know that when they are watching MTV it's like the radio— that you are on heavy rotation, and that people gravitate toward niche areas. The Internet provides a whole other distribution platform that's low-cost, low-barrier entry. Content is an article, a piece of music, a piece of video. But more important, content is now also applications. You can transfer funds from one bank to another. You can pay your bills online. You can use Quicken online. And suddenly content is more involved in your own personal life than just purely receiving information from outside. It is a manipulation of your own information and then the sharing of that information.

When you think about how that works in the context of media properties—suddenly functionality is something people have to think about. You can imagine the consternation of the *Wall Street Journal*, who realizes they have to help people track their personal portfolios as a requirement to delivering financial news. Content, as an application that is consumed and used, is going to be redefined by the medium.

*Old-economy companies not only couldn't innovate quickly enough, they saw their best talent leaving with their best ideas.*

**FERNANDO ESPUELAS**   I was still at AT&T when I embarked on an Internet project, which was sort of a rebel project. It sounds idiotic— you would think AT&T would get the Internet, but they did not. My boss was a very nice man, but he was one of these old dinosaurs who got his emails printed out and wrote them by hand and then his secretary would type them up.

When I came back from the trip with my wife to Nepal, I was changed. It all seemed trivial, what they were talking about. It was this sea of banality; everything they were talking about was exactly what they'd been talking about five weeks earlier. All I kept saying to people was, "It's the Internet. It's the Internet. It's the Internet." Finally, out of desperation, my boss said, "Go do this Net-Inter thing, whatever it is," and he basically told me to get out of the way. And I did.

We launched what was the first portal in Spanish and Portuguese, which was called AT&T Hola. We didn't know it was a portal, but it had the first search engine in Spanish and Portuguese, travel, games, and it had this crazy thing, nine months before Hotmail, which was free Web-based email. AT&T made me take it off the website before the launch, because they thought it would compete with their proprietary email system. I was not successful in explaining why this was so valuable to them, and that's when [StarMedia cofounder] Jack Chen said, "Let's do it on our own."

**SCOTT KURNIT**   If you go to the traditional media companies in town, you'll see a whole bunch of desktop PCs. To me that's a sure sign. When I walk into a company and see a bunch of desktop PCs, I go, No laptops? This company runs five days a week. They've got to lose.

**RUFUS GRISCOM**   Those people who had previously worked in the business world and had experience in publishing and media out-side of academia had the benefit of knowing that there's no such thing as not compromising. From the perspective of launching a magazine, there are a few magazines that have made no compromises, more or less—that have real ambitions. There are two ways to do that: The first is to get a grant and make it an art project. *Harper's* magazine is an art project; they get grants. Then there's something like *The New Yorker,* which loses money every year but makes Si Newhouse feel good about himself. These are all charities.

Coming out of academia, there's something rewarding, though not very idealistic, about the cruelty of the free market. The forces of the free market do not say, We want to pay for you to sit in your attic and paint. That doesn't mean it's not good to sit in your attic and paint. But there's a kind of justice about how the system works, and that's that you have to add value to people's lives. What most people want is to forget about their workday, and that's why most media is really cheesy. That's not an Internet dynamic, that's a media dynamic—although I think that the Internet spirit made possible a greater level of idealism than people would have had otherwise. We could have done Nerve without the Internet; we *would* have. So I think in some ways we're less than pure as an Internet case study. At the same time, the Internet turbo-boosted our endeavor and made it more successful than it would have been otherwise and more interesting.

**ANDREW RASIEJ**   I had to guest lecture an Internet marketing class at NYU, and the first thing I said when I walked in was, "If this class still exists two years from now, the Internet was not a success." It's sort of like saying, "Well, gee, we have fax machines now, so offices with faxes are different than offices without faxes." You have a new tool. It doesn't mean the process of media changes, really. People assume that because of technology a lot of things change, but in reality almost everything stays the same.

*Advertising agencies were especially threatened by the Internet. Accustomed to being the consultant on all branding issues for their clients, they suddenly found interactive agencies at their clients' side, getting them excited about their newfound online strategies.*

**NICK NYHAN** I think the culture we had in Poppe [Tyson] and Bozell was a very interesting case study. An old-school, new-school clash. And it all happened within one address: 40 West Twenty-third Street. There were traditional marketing, research, and advertising people, the martini-lunch folks, who were trying to stay hip but not lose control of this runaway train. Inside Bozell, we saw various attempts to come up with the right formula. First they said, "We're going to have this brand Poppe Tyson, and we're going to funnel all of our interactive work to Poppe.com." Then they were like, "Wait a minute, that's good money, we should do that ourselves. Hmmm. We're going to create another little unit inside of Bozell, and we're going to call it something else. No, we're going to have BozellInteractive.com."

There was a lot of naming, and a lot of management shifts, and you had a lot of antagonism between older people who thought they knew marketing and advertising and these young punks at Poppe.com who were telling the older generation how to market financial services or Circle Cruise Lines online.

There were people on the dot-com side who were arrogant as hell and pissed off the old school, and there were people from the old school who, even if you were the nicest person in the world, resented the hell out of you for getting more attention from IBM, Toshiba, and Dean Witter than they were. All of a sudden they were being upstaged and it was very, very tense. There were joint pitches, but a lot of fighting behind closed doors about how much people should be paid, who should be leading the pitch, and whose advice and strategy should be followed.

It was like Silicon Alley meets "alley." And in alleys there are fights. It was ugly; there was some real bad blood. When left hand and right hand come together, they can either punch each other or they can clasp.

**JOHN YOUNG**   Look at Grey or Ogilvy or Young & Rubicam or us, Tribal DDB, and look at our clients. You can go into any company like IBM or Microsoft, and there's the new media camps and there's the old way of doing things. I used to go down to Bozell and ask the traditional creative director what he was doing and what his schedule was like. He'd say, "Oh, I have to do a TV spot. We do one spot a year." I said, "You have a whole year to do one spot?" "Yeah," he said, "We concept from September to October, then we test from October to February. We go to production February to June. Then we go on vacation." I thought we were walking into Slow-Motion World. We were hyperventilating in this old marketing machine that's been making advertising the same way for the last forty years.

**NICK NYHAN**   It was almost the difference in the length of the carpet—shag carpet down there, and no carpet upstairs. The people who were from the traditional side thought, We throw something up for twenty grand, and if people don't click on it, we'll change it. It was as if this was graffiti on the screen. They'd say, "We don't have to worry about it. You guys are cheap. How much are you paid?" This wasn't like TV production budgets. You didn't need thirty people holding up lights and doing makeup. It wasn't a million-dollar spot, so their attitude was, Just screw around with it and see what works.

**JOHN YOUNG**   And the ad agencies were going, Fuck, we missed the boat, we didn't get it right. We thought it was about advertising, we thought it was about banners.

**MARC SINGER**   I still think the advertising agencies don't get it. I don't mean the Agency.com-like guys who were doing it from the beginning, but the traditional ad agencies.
    A year ago I was in a meeting at one agency, and I was so irate at their stupidity. They were upset that we were telling them banner advertising wasn't everything. Banner advertising's great for some things; people are making a lot of money. But the response rate's not

good enough. It's not going anywhere. They said, "You guys are wrong. Where's your data to back that up?" This was a year ago, and these were top guys at the company. I said, "Look, I didn't invent non-banner advertising."

They said, "Did you tell that to our client?" "Yeah. I said, You might want to do other things." They said, "You had no right to do that. Banner advertising is proven to be effective brand advertising." I said, "You guys are fighting far too hard for this. This is the beginning of something. We're in the early days of the medium and you're locked into your cash cow and you're not smart enough or don't have balls enough to tell your client that there might be other things. There are some other options."

*Because it was easy for anyone to become a publisher on the Internet, offline media companies couldn't decide if the Internet was a threat or an opportunity. One early and eager project was Time Warner's Pathfinder, an online venture that would wrap up all of Time Warner's magazine brands under a single destination site, Pathfinder.com. It has been generally regarded as a major judgment error that TW didn't use its prized brand names instead. The Pathfinder site was also riddled with poor design decisions and a lack of clarity about how best to repurpose content from its offline brands. Founded in 1994, Pathfinder played an important role in the early years of the Alley, but with an estimated $15 million per year in expenses, Time Warner dismantled the project in 1999.*

**JASON CHERVOKAS**   Pathfinder was the whipping boy, there's no question about it. Everybody loved to beat up on Pathfinder. We beat the hell out of Pathfinder, and some of it they deserved. But they did break a lot of ground. We have a myth in America of the first-mover advantage. And the myth comes from Silicon Valley, I think, from technology businesses where owning a patent, or owning something that can't otherwise be replicated, really is a first-mover advantage.

But in the media business and other sorts of businesses, there are many more first-mover disadvantages than advantages, because to blaze trails you've got to do things that people aren't ready for. You soften up the market at great expense and then people come along and learn from your mistakes, and I think Pathfinder was certainly in that position.

But they also were sitting on top of the greatest brands in media, and they were never sure internally whether or not Pathfinder should be a brand. They figured it needed to be, so they could aggregate large enough numbers in a single place to be able to sell advertising. But I think as a result they really underexploited *Sports Illustrated, Time* magazine—great brands at a time when those really could have attracted an audience they couldn't otherwise attract. But that battle went on for years: Do we use our brands, or create a new brand? It was never fully resolved.

**BRIAN HOREY**   The Internet was either going to cannibalize their businesses or slow their growth. It's the reason that IBM didn't dominate the PC business. It's a challenge to the old business and the old way of doing things, and if you want to be successful, you have to destroy your old business in order to succeed at a new one. Companies find it difficult to do that, and don't move very quickly. Their decisions take longer, and a market doesn't sit around and wait for them to move.

What Time Warner is worried about is how their music business is doing this quarter, or how their film or cable TV business is doing. Pathfinder may have been an afterthought, so it didn't get the same resources in terms of money or management. Whereas in an entrepreneurial company the Internet is the sole purpose of the management team and the basis for their funding.

Pathfinder was trying to serve too many masters. They had a portal concept, but it was really Time Warner print mediacentric, which was a different view of the world from what, say, Yahoo had. If you were looking for something beyond Time Warner content, it didn't work very well. It also didn't leverage the unique capabilities of the medium. It was mostly repurposing print content, and the companies that have been most successful have been able to do more than, "We did it in print, so let's do it on the Internet."

**JASON CHERVOKAS**   Time Warner was in a position where it could actually convince national advertisers to take advertising on the Internet at a time where this was a very radical idea. I think Isuzu and Zima were some of the first guys to advertise on the Web. They did it out of their promotion lines, not their advertising lines. Nobody thought they were going to get anything responsive out of this, but thought they'd see what would happen.

Pathfinder is fairly maligned for being a mess. They didn't have clear goals. They had an opportunity to buy a big minority stake in Netscape early on and didn't. It was all shovelware in the early days, and there seemed to be no apparent reason why that should exist. But by the same token, they stuck their neck out, they hired a lot of people, they sold those first ads, they did a lot of trailblazing work that by virtue of the resources at their disposal they could do when others couldn't. I have grudging respect for what they did—with lots of caveats, because it was silly from an actual content perspective.

**FERNANDO ESPUELAS**  When you opened up the browser on Netscape you would always go to the "What's New" button, because every day there would be something cool that was new. Whereas when you would look at things like Pathfinder—if you had used the Internet for more than half an hour in your life—you knew they were totally wrong in their approach. It was fascinating. We kept saying, "What is it that we can see, that they can't see? Why is it that they've designed a product you would never use? It's Time Warner—how can they be so wrong?" The navigation system was incoherent, and it was a fascinatingly ugly product in a world in which design is a big element. It doesn't cost you more to design it in a better way. It just felt like something you didn't want to be on.

**CLAY SHIRKY**  It's normal now to have the Web people not segregated and off to the side. And that was a very strange aspect of the Alley. The Web was so threatening to so many middleman businesses that the typical model was, "Oh I know, let's take the five most famous magazine brands in the world, and let's roll them up under this no-named thing called Pathfinder, and let's spin them out. And we'll just see how much shareholder value we can destroy in three years."

We knew nothing about the Web in '95—we thought people would pay for content—and we still knew Pathfinder was a bad idea. The emperor got up with no clothes on and didn't even think about getting dressed before he started the parade. The only way I came to understand how Pathfinder happened was to understand that there wouldn't have been a Web presence at Time Warner if it hadn't been done that way. The message to the magazines was, Your Web strategy can't be to decide not to use the Web just because you don't like it. So we're going to remove it, and put it over here. It was a tactical mistake.

**ALAN MECKLER**   In my ten years in the Internet field, I would say that there's hardly a traditional media company executive I've met that really understands the Internet. And one of the main problems is the worry of killing the cash cow. If you've got a business that's doing really well, then it's hard for you to say all of a sudden, "Well, we better start investing in this website business," because it might take hold and kill the cash cow. The ones that will be successful, and are starting to be, are the ones that are willing to play both ends against the middle. In other words, to build a website even if it is at the expense of the print, because if you back off on building the website you could end up having nothing.

**DAVID LIU**   Time Warner spent too much money, number one, and they didn't understand what it meant to truly cannibalize yourself. To launch something online potentially threatens the viability of your offline business, particularly when you're still trying to figure out what the online revenue model is. Also, there were not a lot of advertisers, and there were not a lot of consumers using that site.

Think of it from the standpoint of an advertiser who is spending X number of dollars in *Time* magazine. He's a perfect target to advertise in the online version, right? The problem is, if the majority of the traffic of TimeInc.com is being driven by *Time* magazine, and the advertiser is already getting the consumer through the offline magazine, why would he double-pay for the same consumer online?

I think what Time Inc. didn't realize was that you had to actually create new vehicles online to provide value for the advertiser, and ideally something interesting for the consumer when he came to the website. I think that their CNN/SI is an example of when they got it. Suddenly they knew that they had to create a brand that was going to compete with ESPN, and CNN's sports section was not going to do it and SI was too entrenched in its own business model. But they created something else that had a different kind of value for the consumer. And then, guess what—you can sell advertising for it, because it's not really either *Sports Illustrated* or CNN.

**CLAY SHIRKY**   Another amazing story in New York is that the *New York Times* used to be disastrous. They used this stupid typesetting system, which used the "@" symbol as a formatting symbol for type, so they couldn't put email addresses in the text. The sign to the copy editor to change it at the last second to the "@" symbol was to put "at" in brackets—[at]. But the copy editors weren't up on this game yet, so people's email addresses would appear as clayatshirky.com.

**JERRY COLONNA**   The old media companies were always afraid of cannibalizing their existing businesses.

**FRED WILSON**   I'm sure that's true, but I think that a lot of them have gotten religion about the fact that they have to cannibalize their businesses; it's just that they still don't want to spend the money.

**JERRY COLONNA**   The *New York Times* is spending the money. And you know what the difference at the *New York Times* is? Arthur Sulzberger. Arthur Sulzberger Jr. knows that this is the future of the company. He has committed to that, and that mantra has carried through the organization. They are spending the money and making investments for the future of that business.

**SETH GOLDSTEIN**   I think the *Times* has a pretty good franchise; a lot of people depend on it for information. MSNBC.com has been very successful—it's a killer news site. You can't say Time Warner has been successful overall, because they've been successful in some areas and not others. The *Wall Street Journal* has been reasonably successful, but started off with a lot more momentum than they have now. What's clear is that a media company is a media company whether or not the Internet is involved; for a while it felt like media companies were becoming technology companies in terms of how they were valued, and that's just not the case anymore.

**218  DIGITAL HUSTLERS**

**DAVID LIU**  There is an arrogance at Condé Nast that just drips from their publications. I have heard that there is not a single magazine at Condé Nast that is profitable. You wonder how they make their money? Well, they make their money from cable and newspapers. They own a third of Time Warner Cable, and they have a vast expanse of newspapers.

The Newhouses love the magazines, and that's a wonderful thing if you work for them; you get wined and dined and lavish parties and all this kind of stuff. But to try to go online and create a viable business—I don't think they realized the amount of resources that were required to do that. And they weren't really pressured to make money.

When you are not driven to make money you actually begin to fail your consumers, because consumers will tell you what they'll pay for, and the advertisers will tell you what works. And if you are insensitive to all that you'll flounder. People will pay for certain performance, and consumers will demonstrate what they want. To be responsible you have to respond to that, and a lot of these big companies don't do that.

**ALAN MECKLER**  The dumbest idea of all time was probably Pathfinder, but the one that has eclipsed that, now that Pathfinder is gone, is the *Wall Street Journal*. They built the website, but it's a paid [subscription] website. And there are two original tenets of the original Internet that are still around: Nobody wants you to know who they are; and nobody wants to pay for information. And while it's a pyrrhic victory—the *Journal* is often mentioned as very successful; they've got four hundred thousand subscribers—it's a mere pittance of the revenue, traffic, power, and clout that WSJ.com would have if it was free. I think that in another twelve to eighteen months that will be apparent. In fact, I've even noticed recently that they're doing extensive advertising on their radio stations and TV. My feeling is that when you have a big offline property like a print edition, if you have to resort to traditional media to promote a paid subscription or anything on the Internet, you're not in the right business—particularly when it's content and information.

What I like to say—and it's the motto of Internet.com—is: "If you can't sell it on the site, it's not right."

**RUFUS GRISCOM**   When AOL, which barely existed seven years ago, bought Time Warner [in 2000], it was a huge slap in the face—a wake-up call for the entire media community.

**JOSH HARRIS**   I was at this stupid thing up at the Puck building, and there was a woman there who was from—I don't even want to say who it is, but she's basically a groupie. She came up and said, "It's good to see you." I don't even know what to say to her. I don't even want to talk to her. She's still in the new media division of a traditional company. The fact that it's even called "new media" is a bad indicator already. And it's an old traditional company, and she found her spot, and she's in the org chart somewhere. I don't know what to do. I didn't know what to say to her, and I felt kind of bad. I could have gone up to her afterwards and said, "See you later," but I couldn't even gather myself to do that. It was just a ghost. It's like trying to talk to a ghost a lot of the time.

# PURE PLAY

*If there was one thing Silicon Alley indisputably had over the West Coast, it was the party scene. The perpetual motion of New York City nightlife was well suited for the fast-paced striver culture of the Alley. At the height of the boom, there were as many as a dozen parties every single night of the week. After working a twelve-hour day, young entrepreneurs would begin a second shift of socializing. With the open bars and vodka and cigarette sponsorships, party-goers cruised around town from event to event. VIP passes were doled out with deliberate intent, though anyone who knew someone who worked at the company could get on the list.*

*One of the results of all the hoopla was the birth of an Internet culture of celebrity. People began to feel like stars, and popular entrepreneurs were treated accordingly by their peers and by the press. Jason Calacanis, the founder of* Silicon Alley Reporter, *made many of his early Internet contacts while writing a gossip column for* Paper *magazine that tracked the where-abouts and personal lives of the dot.com pioneers. His subjects moved from events like Jaime Levy's CyberSlacker parties to NYNMA's CyberSuds. The Pseudo parties, hosted by Josh Harris, mixed Alley socialites with artists, musicians, dancers, freaks, and hipster kids off the street just looking for some-thing to do.*

*As word of these events started to grow, so did their extravagance and the number of RSVPs. Tracking the blips on the social radar screen was Courtney Pulitzer, a Web designer and young member of the famed Pulitzer clan turned social columnist and later serial party hostess. (To think of the eyebrows and ire Pulitzer must have raised in the family when she purchased www.pulitzer.com for herself!) Pen and pad in hand, she made the rounds of the more exclusive fêtes, taking copious notes and reporting them back to the rest of Silicon Alley in her "Cyber Scene" newsletter, emailed weekly to most of the industry. She then began hosting her own "Cocktails with Courtney" mixers. Donning semiformal dresses, she invited people looking to make business contacts to socialize under the banner of her networking events. The events, always spon-sored, started out as free and open to the public, but eventually included a cover charge. As the media picked up on Pulitzer's story, she spread her parties around the world.*

*Where Courtney reported on the parties that other people weren't invited to, Bernardo Joselevich, a Pulitzer protégé, let everyone know what parties were coming up and available to the growing masses of people in the burgeon-ing industry. An unproven entrepreneur (DutyFreeGuide.com) and recent émigré to the city by way of Israel and Argentina, Bernardo followed suit not by hosting his own parties, but by announcing everyone else's. He cornered the market, and his party list proliferated, even starting to create some revenue as Alley companies placed help-wanted ads in the body of his party updates. And for a while it worked.*

*As often as not, an Internet party RSVP page would be set up online at the company's name with the extension "/party"—www.doubleclick.com/party.html, for instance—and any party crasher could reply.*

*In late 1999 and early 2000, the pace of parties reached a frenzied pitch. But when companies began to lose their footing, and the future became more unpre-dictable, the number of events dwindled. Soon lists such as Bernardo's featured more entries for business seminars than launch parties.*

**JAIME LEVY** I had my CyberSlacker parties [in '95], bringing together animators and programmers. It was a party atmosphere, but we would have computers set up. People got the opportunity to know each other. It was quite good for me, because basically everyone in town knew me back then. I didn't realize how much of an influence it would have; lots of people know me from that.

**CLAY SHIRKY**    When Jaime Levy came along, she had graduated from ITP and built Malice Palace, which was an incredibly weird dystopian environment. Jaime became one of those people for a few years who everybody knew. If two people knew Jaime when they met, there was something to talk about.

**JAIME LEVY**    Josh Harris came to one of my parties and said, "Wow, these Internet people are really kind of crazy." But he doesn't admit that the Pseudo parties were influenced at all by the CyberSlacker parties.

**JOSH HARRIS**    The biggest party I ever did [at Pseudo] was with Gen Art, when they were first starting. Well, we got them launched. I'd just taken over the fifth floor [at 600 Broadway], and it was empty. So I said, take the fifth floor, have your party.

Meanwhile, on the sixth floor, I'd found this other group from downtown. And I decided, I'm going to run a real casino. So I rented some blackjack tables, had real dealing going on, but didn't really have a big bankroll to back it. So we weren't doing too well with the whole thing.

Then madness broke loose, because word got out, and the whole building basically filled up. This was the only party I did that was truly out of control and truly possibly dangerous. I'm sure fifteen thousand people were in that building at one point or another. It was packed. People were just coming and going all night, and by the end—this was before Giuliani caught on to me—I had a traffic jam of police department and fire department vehicles blocking off all the streets on that corner [Broadway and Houston, one of the busiest corners in New York City]. And they still let me go, because that was the way New York was run back then. Nobody got hurt, and whatever.

**JAIME LEVY**    After me, the Pseudo parties started happening, and that other stupid one—CyberSuds—which was really like the corporate, lame version of my parties. They were just full of men and really boring.

**CECILIA PAGKALINAWAN**   In 1994, I was a media specialist at Young & Rubicam. I was trying to push this whole Internet thing and said we should offer services and so forth. From the powers that be, the response was: "This is a fad. This is going to go away. So let's ignore it."

So I would go to CyberSuds and be one of twenty-five or fifty people and one of three women there. A lot of us were drawn to each other because we were the only ones who really believed this was going to grow. There was definitely a very tight community feeling, because we were the only ones who actually thought each other was sane. All our families and our friends and the companies we were working for thought we were crazy.

**STEFANIE SYMAN**   The entrepreneurs would be hovering near the food tables and the bar for the free stuff, and the suits would be mingling and trying to do deals or whatever. We'd be kind of comparing notes on the scruffy, scrappy, trying-to-make-it world.

**SETH GOLDSTEIN**   It wasn't just parties for the sake of partying. It was a group of people who were all taking real risks, and early on there was a real appreciation for innovative thinking. The kind of website you built, the kind of banners you did, the kind of media you bought was judged on the quality of the idea. As it became a machine and companies like DoubleClick became a real business—potentially; time will tell—it was no longer about groping in the darkness.

**JOHN BORTHWICK**   We did a lot of events, which was the early way we actually could make money on the medium. Because back in those days people were willing to spend discretionary PR budgets on getting involved in Internet things, because it seemed like it was the hot thing. So we did, like, five or six events—big events.

One was called Cosmic Cabin. There was a room at the Tunnel which was all done up like a '70s Kenny Scharf design, and we created a virtual version of it. We then put computers in the real nightclub, so you could sit in the real nightclub and go into the virtual version and chat with people in the virtual version—basically bringing together real and virtual space.

**ALICE RODD O'ROURKE**   Networking has always been important in Silicon Alley. I've worked in four industries, and I've never seen networking the way this industry networks. It's very purposeful: "Don't you want to know? Don't you want to share? Don't you want to be a part of this?" It's got a spark in it that is very exciting.

**MARISA BOWE**   We had strippers at our second anniversary party down at this art space in the Wall Street area. We hired a DJ who also screened porn flicks behind him. One of the people who worked with us—an editor at Charged [Word's sister magazine at Icon]—was a gay male go-go dancer at Pyramid. He brought along two of his friends. I also had seen someone at Thread Waxing Space who was a major Chelsea go-go dancer, and he brought some friends and performed.

Our parent company, Icon, was filled with all these computer-salesmen-from-New-Jersey types. The hipsters weren't so shocked, but to the computer people it was like going to Scores [a Manhattan strip club]. For us, it was about celebrating Word and our independent culture.

So I got this phone call some time later from this reporter at *Alley Cat News*, saying, "Do you think having this type of entertainment is really good for your business?"

**ANNA WHEATLEY**   We wrote a feature about how having strippers was not such a good idea. I worked for Bob Guccione, so it's not like I come from a background of squeamishness, but it's that hip and trendy thing that bothers me. In terms of the content on Word, she could do whatever she wanted, but when you're hosting a business function for your company and your employees, you do have a responsibility. Can you imagine if that had happened at IBM? Being an entrepreneur sometimes seemed like it had nothing to do with being in business. Being an entrepreneur was great, cutting-edge. It was like having to make money was something terrible.

**MARISA BOWE**   They just couldn't understand that we were a media business, not some database company trying to sell ourselves to major bank clients on our stability. The dancers were perfect with our demographic. They were the sexiest people that any of us had ever seen in our lives.

**JASON MCCABE CALACANIS** We had party pictures in the back of the *Silicon Alley Reporter.* It was the big thing: who was in the party picture, what were they doing? Austin Bunn was at the Word party with some cross-dresser, grabbing her fake bosom. A very controversial shot—Jeff Dachis putting a dollar bill in a G-string of a, you know, dancer at the Word party; we always had sort of risqué pictures in there. Things that could never happen these days. Jerry Colonna kissing another guy on the cheek, all these splashy photographs. It was just good fun.

**JASON CHERVOKAS** It was very much a tight-knit scene. For the first year and a half after we launched @NY, Tom and I corresponded with a small clique of people who were trying to figure out what was going on. It was very personal, it was very intense, it was very collegial. We sensed that we were really experimenting with something new, and we were just kicking ideas around. Everyone knew everyone else.

**RUFUS GRISCOM** We threw our first party for nine hundred dollars, and we found a bar that wanted exposure down on Orchard Street. Orchard Street was just kind of emerging then, but it was nothing like it is now.

You would occasionally have these moments where you felt like you were participating in a cultural moment. But you only had that head-above-water perspective once in a while. Usually you were just kind of like, "Well, I'm living my life," but you didn't see it in a larger historical perspective. You didn't see it as a sort of sepia-toned, "early Internet experience."

We were a couple of young kids with not much money who were doing something that was fun.

**JOHN YOUNG** There was one Razorfish party, one of the first ones, that was sponsored by their vodka client. I remember it being just this huge pyramid of vodka in the middle of this big abandoned space, and they had belly dancers and all this stuff, and there was all this email controversy on the WWWAC list afterward about whether it was politically correct to have belly dancers.

**CLAY SHIRKY** I can think of few other things that have had the same effect as the Silicon Alley 100 list becoming an institution. The Razorfish parties, I guess, and on some level the Pseudo parties, though they were more low-key. At one Razorfish party there were people standing in line to get in. It was a booze-fed bacchanal, with professional go-go dancers, and there was this huge thing the next day about this being sexist. There was a big crisis between the Alley's image of being a kind of new workplace, free of some of the issues of older places, and then Razorfish goes and has this party that looks like a Hollywood blowout, and whether you were at the party or not had some of the same effect as whether you were on the list or knew people on the list.

**NICK NYHAN** Kevin [O'Connor] was one of the first people to have weekly cocktail parties. He knew that the Internet was being run by young, single people. So he went out and sponsored all these parties at bars because he knew that they all want to meet each other.

**LISA NAPOLI** I would go to parties every night, and I loved it. It was just such a fine time. This was before it was all about formal parties, money, all about press agents. It was a very informal kind of world. There were all these Internet cafés at the time. It just had a sense of excitement.

Everybody likened it to the early days of TV, and you got the sense from all the people who would keep showing up at these events that they knew, you knew, that you were onto something. No one knew quite where it would lead—you had a secret that was just becoming public—and it was just a thrill to go to events. New York is such a huge place, but a small place in an odd way, too. It was nice to know there were people who had the same vision about what was happening in the world as you, and were working on it somehow, and that you could come together and sort of commiserate, even if you didn't really talk directly about it.

Of course, in the early days CyberSuds were great events. I got into CyberSuds later than when it was ultracool, just a small group of people at a kind of salon. It was much bigger by the time I got to it, and people really ranked on it after eight or ten months, because it got so enormous it became like an enormous networking party rather than just a place where people hung out and had a beer and commiserated after a long day.

**COURTNEY PULITZER** | COCKTAILS WITH COURTNEY | As I started getting out there as a designer, doing more websites, one of the sites I did as a freelance job—for free in exchange for a link and a little publicity—was for @NY, which was just a small online publication. I always kept in touch with Tom and Jason, did a little graphics work for them here and there, and then about a year after doing their website, they said, "Hey, would you like to take over the social column?"

I'd never written a social column for an online publication before, so it was sort of a new concept. But I said let's give it a four-week trial, see how it goes. The first column was in April of '97, and the trial never ended. It was so inspiring and amazing to keep getting out there. And now, with pen and paper, to be able to chronicle what was happening in the industry. Each week was really fantastic.

That column appeared every other week; then I went weekly. And after writing for them for about a year and a half or so—also for free—I thought, Gosh, you know, I might want to do this on my own. I didn't really skip a beat. I just continued publishing my newsletter. Now I sent it out, just on my own list, and over the years with the cards I've been collecting, I started anew with my distribution list. So that's how I kept moving on with the Cyber Scene newsletter.

**MARISA BOWE**    It was fun at first to see all the groups of people we had something in common with. Then it just mushroomed so suddenly, seemingly overnight. It was kind of surreal.

**SYL TANG**    DoubleClick had this "Willy Wonka and the Chocolate Factory" party. They had gigantic mushrooms in psychedelic colors, people dressed the part, and the favors were all Oompa-Loompa things. The tickets were "Golden Tickets," with a raffle for a golden candy bar. The invitation to the party was a little golden ticket. It was an extravagant use of money. It must have cost a fortune.

**KEVIN RYAN**  The Willy Wonka–themed party was a large one; it was the first time we had twelve hundred people. But we do a lot of smaller events. We had a great wine tasting event out here [points to terrace garden], obviously a completely different type of theme. We're doing a charity basketball event here. So it's a real variety of things that change depending on the location and the target audience. We do a lot of smaller, what we call "off the record" dinners, where we invite CEOs or senior-level people—only twenty people at a time—to a great restaurant, where we talk about the future of Internet and advertising. A totally different type of segmentation, but nicely done. We have great people in events planning. We do spend some time on it, and it's part of marketing, and that can be as effective as, or more effective than, buying an ad.

**MARISA BOWE**  One of the first Super CyberSuds events was held at Astoria Studios or something, in this huge building. I ran into Omar Wasow, whom I knew slightly because we were all in this little group. I noticed there were all these kind of slick-looking people in suits handing out business cards. I figured they were consulting about the Internet, but they didn't know anything about it. They were charging much more money than I would ever dare to charge for this. Omar and I just looked at each other, and we thought the same thing at the same time: "What's up? Who are these people? What is this?" It was so weird.

**JOHN YOUNG**  You went into CyberSuds with excitement and fear, because you'd always get pinned into a corner with someone with a lamebrain idea who would just talk your ear off. Now I can't even keep track of the parties. I can't even bring myself to read Bernardo's [Joselevich] emails anymore, because I can't go to the parties.

**JOSH HARRIS**  I guess I'm among a group of people who are sort of snobs. It's not even that. It doesn't happen so much anymore, but it gets to a point where I don't want to go to public events because I get attacked. It's like, I don't want your business card. If I wanted your business card I'd ask. Then it's like I'm rude or something. Then I'm thinking, That's why everybody's here—so I don't really want to be here. I don't need to do business here, and these are not the people I want to hang around at this point in my life.

**SYL TANG**   Silicon Alley was a place where you could go up to anybody and talk to them for no reason whatsoever; you could go to any party and have an excuse to talk to any person.

The social scene in Silicon Alley used to be one where you would go out to parties and deals would get done. Because it's a lot of young people controlling a lot of power and a lot of money, they would go out and put these deals together. Deals were being done for all kinds of crazy reasons. You would go out with someone and have a drink and find some random reason to do a deal with them, for no other reason than you liked the people and it would be cool to work with them. There were more parties than you could possibly go to at any time.

**BEN SILVERMAN**   When you have a lot of money floating around, when you have a lot of young people, when you have a lot of stress, the drugs are always going to be there.

Coke is pretty popular. The party drugs are definitely popular. I don't do the clubs a lot, but if you go to certain clubs, I always see people from the industry there, and I know they're on drugs or using drugs, even selling. I think that's as much a part of New York culture and downtown culture as the industry itself. Coke is definitely popular because it spills over from the Wall Street crowd. Cocaine is just generally popular right now in New York.

**RUFUS GRISCOM**   I remember realizing for the first time that someone thought that fooling around with me would be good for their career. I was like, Oh my God, you gotta be kidding me.

**COURTNEY PULITZER**   It's the one industry I've seen where networking is such a vital, major element to its growth, and to its development for a few reasons.

One, especially in New York City, this is a town where everyone goes out to drinks after work with their colleagues, business partners, et cetera. It's a very social city to begin with, more than many others.

And then the industry is also fed by waves of young people who graduate each year from school, who are also very social people. They're always trying to get out and hang out with their peers, meet potential boyfriends and girlfriends. Just because it was so youthful, and then because of being in New York—and also because it was so small in the beginning—people were trying to meet up with as many people as they could, to get a sense of justification that this was a really valid thing, that the Internet wasn't going to go away, that it was going to be a really big business.

And now, of course, we all know it's a big thing—except now it's about staying in business. Some of the justification has changed. The industry is changing at warp speed, so it's a great way to keep in touch with what's going on. Sure, we could all get newsletters, but when you're out getting to meet with people—especially last year, when there was so much noise with everything—unless you were out, you would get lost in the shuffle, be just another price-comparison website. But if you were out at the parties, and people got to know you, then maybe they would be that much more willing to deal with you instead of your competitors.

**SYL TANG**  The DealTime party at the Barney's building was very lavish. They gave out DealTime martini shakers, which will probably become a relic. They had these people going around telling your fortune. Everybody had a T-shirt on that told you what they did, so there was a massage therapist, psychic, palm reader, matchmaker, all sorts of things. There were two people with T-shirts that said "photographer," and they would take your picture and give it to you. They spent a lot of money.

**COURTNEY PULITZER**  I started the "Cocktails with Courtney" events in '98. They were really just an attempt to get together with my friends in the industry, because I was running around to four parties a night, seeing all my friends, but never really getting to talk to them at parties. You always have to write about new people, and if you write about the same five people all the time it's not really interesting, it's not good reporting. So I just said, That's it. At least one night a month I'm going to stay put and stay in, invite my friends to come see me. It's like the theme song to *Cheers* playing in my head. You know, it'll be a place where everybody you know gathers, the last Wednesday of the month; everyone will know your name, and whatever other parties come and go there will always be Courtney's party.

I figured I would call it "Cocktails with Courtney," and get fun little napkins. And my goal was to do one a month for the next year, and just see what happened. I just wanted to do that, complete it. And after doing a few—well, the first one actually took off phenomenally, more than I needed it or wanted it to.

We've had a huge response, and it's never slowed down. Because people knew about it from all around the world, and all around the nation, especially. I started getting all these requests, saying, "Hey, when are you going to come out to the West?" or, "Don't ignore us in Austin." And I started looking at these requests and figured, Okay, if you're interested in sponsoring, then we'll do one there. And last month we had six cocktail parties in six different cities, from London to L.A., and everywhere in between. So it's definitely taken off in an amazing way.

**JASON MCCABE CALACANIS** [Courtney Pulitzer] is the weirdest thing to come out of this whole mix. I think she's harmless, I think she probably has very good intentions—but bizarre, just absolutely, stunningly bizarre. Courtney's always been kind of tough on me. She goes to my parties and writes nasty things about them, because everybody looks at me as competitive or arrogant or whatever. My parties are the best parties you could possibly go to, and then she pans them or doesn't cover them.

There are a lot of fringe people. If you go to the indie film world, there are fringe people, like hacks, people who aren't part of the industry but who sort of glom on. They become part of it, but they're like these weird people. And I think she's part of that weird group of people that wanted to be involved in the excitement.

I guess I was the original Courtney Pulitzer when I did those pictures. But I realized there was a limit to it, and I didn't want to be pinned as that. God forbid I got pinned as the Courtney Pulitzer—I'd kill myself. She's getting beat up a little bit in the news [now]; people realize it's like a really sort of weird thing, a socialite in Silicon Alley. She's a wacky bird, that's all I can say.

**SYL TANG**  I met someone last year at a networking event who wanted to get into Silicon Alley, for all the wrong reasons—for purely financial reasons. We started dating, and we got very serious very quickly because we had a lot of similar interests.

One day we went to a Cocktails with Courtney, which at that point was not this gigantic, international affair where you had to register to go to Ohm. It was thirty people who got together in the back room of Bowery Bar, all of whom knew each other, but none of whom knew Courtney. We met a woman there who was a VP of Biz/Dev at a finance site. She chased me very hard to do a deal with my site, which I thought was a little weird because at the time we hadn't actually launched. We negotiated a contract over the course of a month. The Friday that I received the contract and was supposed to sign it, I was swamped because I was flying to Australia. I told her I would get it to her in a week when I got back.

While I was away I found out via email that she and my boyfriend had in fact been sleeping together for the entire month—from the night we met her. Literally: We met her, we went to dinner, I went home, and he went and slept with her. The same Friday I was flying away to Australia, I found out that he had been at Van Cleef & Arpels looking for engagement rings for me with his best friend.

What they both did was incredibly sleazy. She used the guise of the deal to find out my schedule, my life, the inner workings of what we did. The night I met her at Cocktails with Courtney, I thought she was a little bit strange, because she was very interested in me. She was like, "I love your clothes. Where'd you get your boots?" She was touching me. I've never had a total stranger start feeling me up. She was just very into me, and she wanted to know a lot about me—evidently to the point where she wanted what was then my boyfriend, maybe fiancé.

So when I came back and found out, I killed the deal and dumped my boyfriend.

**BERNARDO JOSELEVICH | BERNARDO'S LIST |** My first list was actually an email to the guys that started Favemail, which then was called ClickMail—a company that got funding from Flatiron Partners. The founder came from the Midwest, and he and his wife wanted to have things to do.

I offered to start sending them invites, and I started aggregating them, and they started forwarding them to friends, and then the friends said they would like to get them directly without having to rely on their forwarding them. I guess the list officially started when I put a headline saying, "You're welcome to forward this to other friends in the industry."

I was always careful to make it completely open, from the point of view that nobody would get it if they didn't ask for it, and that no private event would be listed if the host hadn't requested or given their blessing. It generates a lot of goodwill when you're respectful of either people's privacy, or their right to invite whoever they want to their parties.

**BEN SILVERMAN**   Now the industry parties are hilarious, and the people are just sad. I call it the account executive syndrome. Everyone's an account executive, which means—not to be mean, but you're nothing, basically a salesperson or whatever, just handling projects. Of course you want to move up from that, so you'll kiss anybody's ass to do it.

The industry parties are boring as shit, because it's the same people and nobody is terribly exciting. Everybody's looking to make a deal, and it's an everybody's-selling, nobody's-buying mentality.

*Josh Harris once said he believed he was put on earth to throw parties. From 1995 to 1999, his raucous Pseudo parties more than served that mission. But nothing could compare to Quiet, the party he threw to celebrate New Year's Eve 2000. It was a month-long affair that took up two three-story former textile mills in the fabric district on lower Broadway.*

*It was the peak of the boom in Silicon Alley, and Harris spent his money with thrilling abandon, burning more than a million dollars on the event. He invited artists and friends from around the world to become "Citizens" of his premillennial community. Citizens lived with him inside rows of pods, steel beds connected to digital video cameras, from which they could watch the action in each of the other pods and everyone else could watch them on a closed circuit.*

*For weeks, Harris threw a continuous stream of miniparties and events in the Quiet space, mingling the Citizens with anyone savvy enough to find out the address. In the* New York Post, *a few days before New Year's Eve, he made an open invitation to anyone to attend.*

**JOSH HARRIS** At this point, I was at one with the tune of the city. I was making a film. I just showed up one day and started doing this stuff. There's a lag time before they [the city] figure out something's going on.

The local police really never liked the mayor, so unless he says to do something they really would rather not. I had a three-day party in the fall of last year, and then for Quiet I was in the same space for a month.

I kept it just below, right up to the edge the entire time. I knew they didn't want to hassle with it, there wasn't enough of a problem.

There was a lot of activity in the pods, and maybe that's a part of the genesis of this place [WeLiveInPublic.com, Harris's post-Pseudo project]. I started realizing that people wanted to be seen on camera, and I kind of got a sense of the dynamic.

One night JUDGECAL from Pseudo starts waking everyone up, saying "Go to Channel thirty-six." Three bunks over is Channel thirty-six. And I thought it was David Leslie [a performance artist and friend of Josh], but it turns out to have been a professional plant. This is the depth of things. He planted somebody, in his bunk with some other woman, and they were going at it—which was okay, but they were going at it, must have been an hour and a half straight. Literally an hour and a half straight. But you're watching them, and you're hearing them, and that sort of set everything off. Everyone said "Aw, fuck it." You start tuning to different channels, and you could see them and you could hear them—and the same with the shower, which was open, and the toilets, which were open. It created a sort of intimacy which actually turned into a cult. Now I know how to make a cult.

So on New Year's Eve I was making this film, and I got to make my opera complete. In the front windows I had Maya Hanson, who's a fashion designer, put very scantily clad women with a red background on trapezes, and I put up a big sign that said GIRLS, GIRLS, GIRLS, XXX, blinking, right in front of the courthouse.

So I put my sign up on New Year's Eve, just at twilight. Of course, I knew the whole city was on alert for all kinds of other more important things, so I put it up and made sure that all the cops going by, and the mayor, were going to see that sign. It was going to get to the mayor's office.

We had a beautiful New Year's. No problems. Actually, I had *some* problems. I used the police—the local police were great. I called the cops over, said "Get these guys out of here," and they took care of business.

The stroke of midnight at New Year's Eve 2000 was not the best time. I was actually supposed to do simultaneous orgasms, which I'd done the previous week—a few days before? Can't even remember; it's all a mush to me—and it was like, We did it! And I was going to do it again, I was all set up, and it didn't feel right. I didn't feel good about it. What I ended up doing is gathering everyone, all of the Citizens, together in the pod room, and we just sat there and took a photograph, and did a little video.

Just at the stroke of midnight, everybody was quiet. We just sat there. All the young girls wanted to pop the bottles, and it was like, No, hold still, just stay, stay, we've got video rolling and we want to keep it there. And that seemed to be the right call.

Then we had this weird thing happen. We had the Feast going and whatever, but we really didn't invite anybody. There were no invites. People just showed up. There was the art thing upstairs, and I locked them out. I didn't want to hang around them. So I had all these people milling around upstairs who didn't really know what to do. It was like, I didn't owe them their New Year's—that wasn't my responsibility. I felt a little guilty, but heck, we had these weird people milling around upstairs.

Midnight came and went; it happened; we got our shots. And then about one o'clock or one thirty, it got cool. It found the groove. We were kind of done. And then the art world showed up, all the dealers and all the artists, and we just sat in that room from one thirty in the morning till about six-thirty.

But just like clockwork, after they cleaned up Times Square and cleared out, I had the whole security forces of the United States government, including New York City, all there at Quiet. They gathered at the end of the Times Square gig, and they had their after-party at my place. I had the captain of the police force, captain of the fire department, a

SWAT team, the local fire department, the local police department, and the Federal Emergency Management, the FEMA guys—it must have been a hundred vehicles and five hundred men, breaking down my door in the early morning with nothing going on. I mean, everyone was sleeping in these bunks.

There was the machine gun room downstairs that they had to go in. This all made the news—I mean, I'd put the party on the Fox news that night, so they knew when they broke in that there was nothing going on. Again, I just had no respect for what they're doing or what they're about.

They didn't break into my place for the right reasons. They could have knocked on the door; they could have brought one guy there and said, Hey, you guys have to go. On the other hand, it was an after-party, and they were having a good time. I have pictures of them with smiles on their faces, 'cause they all knew—they all knew they had no respect for the system.

Of course the videographer woke up, I got tape and photographs of all these guys breaking in, and now I have the end of the movie. I have the operatic end of my movie. I knew we'd do it. They thought they got me.

The day the fire department came it was punctuated. It ended. Fine by me. But at the end, for two or three days, all of the people who were Citizens didn't know what to do with themselves, they were lingering outside in their uniforms. But we wouldn't let them in because it was over.

I spent over a million bucks on Quiet. I'm telling you, I don't know how long I'm going to last. My view is, I'm in my prime. I'm pretty simple. I've never been a money guy, so we'll see what happens.

The people that don't make it generally take too many drugs and burn themselves out. I've mastered the conservation of energy. Though I have to say, I was sick for six months after Quiet. Not physically, but sort of mentally, I was sick.

I'm good at figuring out how to rejuvenate myself, because in order to do this stuff you have to be willing to go right to the edge and stay there. Quiet was on the edge for thirty days. And it wasn't till I took a month off and went to Africa—a painful month overseas—that I came back and was basically good to go. It was a kind of withdrawal or decompression or unwinding or something. It hurts; it's mentally painful. You just kind of do your time, and once you're back to health you don't even have time to enjoy your health.

*In the weeks leading up to New Year's Eve 2000, there was a widely held concern that New York City would be victimized by terrorism, or at least a bad case of the Y2K virus. When people awoke in the next few days to find the world mostly unchanged, there was great relief, but also, perhaps, the first sense of exhaustion at the years that had passed at Internet speed.*

*The NASDAQ continued to rise for the next few months, but there would never again be such wild, innocent, hopeful frivolity. By the end of the following summer, Pseudo.com, the party Josh Harris had thrown for over five full years, would be gone—and with it hundreds of other businesses, thousands of workers, and the innocence of the early years of the Internet industry.*

# IT WAS ALL A
# PSEUDO DREAM . . .

*J*upiter Communications, an online financial analysis company, could be called the Alley's first highly successful venture. Josh Harris, then twenty-seven and working out of his apartment, started Jupiter in 1986. When it went public years later, it returned hundreds of millions of dollars to its investors and founding partners.

When Josh found himself being bought out of his position in Jupiter in 1995, he jumped at the opportunity, starting a new company called Jupiter Interactive. Before long it would be relaunched as a Web broadcasting company, Pseudo.com—early enough in Internet history that words from the dictionary were still available as Internet addresses.

Investors knew they had made money with Josh before, and felt he was a good founder—he knew when to start out and let someone else take over. He may have been quirky, but who could argue with success?

Pseudo's first victory came when it won the chance to run Prodigy's Chat application in return for a percentage of the profits. Harris used the windfall to begin to build a radio program, then an online radio website, then a full-fledged *interactive broadcasting site, with fifty hours of original programming on a wide range of offbeat subjects.*

*People often said Pseudo was ahead of its time, that its curse was being too early. Pseudo was one of the first interactive broadcasters, but plenty of experiments in Internet television had failed throughout the '90s.*

*Pseudo's online content was divided into a variety of separate channels with a few programs on each channel. Each channel featured a different genre, such as music, sex, gaming, politics, science, and more.*

*The programs were designed to appeal to niche groups. Desivibe, for example, which covered the Bhangra scene in London and the United States, was probably one of the top news outlets for South Asian music lovers. The shows on CherryBomb, including the Women's Channel, were extremely forward-thinking and raw.*

*But since most of the site's video content required that the user have a broadband connection, the potential audience for Pseudo's content was a fraction of the general Web population. Most of the people in the Alley expected the bandwidth to become widespread much more quickly and overestimated what the climate would be like in the immediate future. Pseudo used digital video, which it edited, compressed, and then streamed over the Internet. The most crucial problem was that broadband access to the Internet— anything over 56K—was available to only a fraction of the country even until the end of 2000. Most Pseudo visitors were other dot-com employees on T1 lines or students connecting over university networks. Still, it was an audience with the potential to become "early adopters" of what had the potential to become the world's first new media brand. Brand building is the entrepreneur's ultimate holy grail—because an MTV or CNN will not only return money again and again, but become a part of the culture and change the culture around it.*

*Many people, too, complained about the "public access" nature of the shows, comparing them to the ragged production values of unregulated community television. Josh genuinely felt that if the Web was going to be a broadcast entertainment media, its hit shows would have to be invented under the new rules, and he intended to create an environment that would make the medium work.*

*Throughout the late 1990s, many bright people in New York thought Pseudo would pull it off. The sale of Broadcast.com, an early online audio pio-*

*neer, for 5.6 billion dollars to Yahoo in April 1999 only seemed to validate the notion. There was a sense among the workers at Pseudo that an IPO would turn every street kid Josh couldn't say no to into a millionaire.*

*Every time Pseudo seemed on the verge of running out of money, Harris and his team turned up a new slate of last-minute investors, bringing in serious players like the Tribune Co., Intel, and Prospect Street Ventures, a respected New York VC fund. All together, Harris raised over thirty million dollars to fund the company and had more than two hundred people working out of three offices housed in prime real estate around Broadway and Houston Street, in the northeast corner of SoHo.*

*The main Pseudo building, at 600 Broadway (recognizable from afar by the enormous DKNY painting of the city on its side), became Harris's central base of operations. He took over one of the floors as his personal loft and built colorful offices for the several channels of the network. Staffers became used to seeing their eccentric boss dressed up as Lovey, his clown alter-ego.*

**GENE DEROSE**  I think I'm one of those rare people who had a clear view of Josh right from the outset, and I don't think I really changed it after the first week I met him. One of the most unorthodox people I'd ever met. He was also, though, the first entrepreneur and business owner I'd ever dealt with at any close level, so a totally mixed combination of very quirky person and, at least at first, a business authority figure for me.

He was very focused and intelligent when he wanted to be, but had very far-ranging sets of interest and creativity, not ever wanting to really get pinned down in any one place.

**JOSH HARRIS**  I put an ad in the *New York Press*. Gene showed up, and he was almost like just a warm body. I didn't pay him very much. This must have been 1988 or something, and we just sort of worked things out over the years. We went two or three years trying to figure out our deal. There was nothing written on paper; we just sort of figured it out. And he was hungry for it, and I knew I wanted to do Pseudo, so we made our plan, me and him, and brought in Kurt Abrahamson, Gene's old-time friend.

And then the market started to hit. I count my lucky stars for having

Gene show up, because it worked out. He was born to the business, much more than I was. Actually, by the time he showed up I was already doing it for five years in one form or another, so by the time I left Jupiter I'd been doing it for eight years or something. I had kind of done my time. I think I left on top.

When I left, he and I were starting to fight over who got quoted in the press—not really fight, but it would be a competitive thing to see who would get the quote. I knew he knew what he was doing, and it was very comforting. As the thing was starting to pop, the *Wall Street Journal* and the *New York Times, USA Today,* and the *Washington Post* were covering us, and where you were placed in the articles was the litmus test. And I was starting to be placed very consistently at the right spot in each of the articles. To me that was the indicator that you're on top of your game.

So there was a point in time—I would say for two or three years— where I was the best in the field. Maybe the world at that time didn't acknowledge it, but I was the best in the field. And it was like, All right, now what do I do? And then Gene, as it turned out, became the best in the field. I like to think that he was well-trained, and then he proved out that he was not only the best in the field but also a superior businessman.

**GENE DEROSE**  Josh never had a straightforward sense of conventional business. There were never really any propriety or illegality issues per se, but when he left Jupiter to start Pseudo, he should never have called it Jupiter Interactive. For some odd reason I wasn't thinking through the implications at the time; we let it just happen that way. From day one, from the day he left, they had nothing to do with each other. Their economics were separated out. Kurt Abrahamson had started at Jupiter in early '94; the only remaining tether was that Josh owned a chunk of Jupiter. Within a year or so he changed the name to Pseudo.

*The company cut a crucial deal with Prodigy to share revenues on the Adult chat application.*

**JOSH HARRIS** Scott Kurnit basically did the deal. It was him. It wasn't the other lame guys at Prodigy. When the checks came at the end of the month, we were doing twenty-five percent of all traffic on the Prodigy services company, which at the time was the largest online services company in the world. Right? So whatever we did with them worked, and he was the one who brought us in. They were making—add overhead and all that, but they were making ninety cents on the dollar for whatever traffic we brought in, and we were making ten cents on the dollar. They were doing pretty good, and we were doing pretty good.

What happened was that he left, Bill Day left, Catherine Hickey left—they were the key people there—and Prodigy got sold, and then Ed Bennett came in. While Ed Bennett was there he sold the company to this guy up in Boston, who came in with an ego, and was not a practical businessman. He was a good businessman, but his ego was involved; it was too big for him. In my opinion.

**NICHOLAS BUTTERWORTH** Josh Harris—this shows you how smart Josh is—took over a whole wing of the building at Prodigy's White Plains headquarters and, legend has it, had his own key made so the Prodigy execs couldn't get in to the space he had taken over. Not only had he commandeered their space and their engineers, they couldn't even get in to see what he was working on.

Josh was brilliant at getting the concession to do adult chat on Prodigy, and whoever let him do it was clearly an idiot who didn't understand the dynamics of online service businesses. AOL, of course, was built on adult chat; Yahoo was built on adult searching. If you took all the adult content off the Internet you wouldn't have a whole lot left. Josh had the only area on Prodigy where you could use free language. Everywhere else the language was censored and the topics were censored. He said, it's going to have a lot of topics, including art and fashion, and meeting people, but it was the only place you could get hard-core sex on Prodigy, and he was getting a cut of the connect-time revenues. Obviously sex chat is the biggest driver of connect time after email, so he cleaned up. So in that sense I think he was very smart.

**ROBERT GALINSKY** | EXECUTIVE PRODUCER, PSEUDO.COM |
Prior to October of '95, I didn't own a computer. I was heavily into the spoken-word scene, doing a lot of theatrical spoken-word perfor-mances. I met Josh in early '95. Honestly, I didn't really like him right away. I thought he was exactly what he is, which is pushy and arrogant and sort of a know-it-all. At the time, I didn't know he could back up his know-it-allness.

I was producing a show called "Galinsky's Full Frontal Theater" on Twenty-eighth Street. It was a Thursday night show, an all-night open mike. At midnight we would start serving pasta and wine for free. Josh used to show up at midnight with a bunch of people with weird instru-ments made out of trash and garbage. I had heard he was an interesting dude, so I said, If you ever have a project you want to do, let me know.

First thing he pitched to me was "Let's do a book—a coffee-table book about the Internet. We'll get screen grabs and we'll do interviews with people and we'll have a cool art book about the net." So I bought a couple of books about how to write a book and spent a couple of months trying to figure it out and then just called him up and said, "Look, this isn't the right one. Something else is going to have to work before this one."

He called me back about a month later and said, "We want to do a radio show about the Internet." And I said, "I know nothing about the Internet. I don't own a computer. I'm not into it." He said, "It doesn't matter. You know how to produce and keep people interested. And that's what I need. The rest I can teach you."

*Josh seemed to trust people, consistently, naïvely, and blissfully, to live up to their potential.*

**ROBERT GALINSKY**  Prodigy chat was netting us a hundred and fifty thousand dollars a month. And I used to watch Josh late at night clicking through, and counting how many people were in the chat rooms, and doing the math: percentage of people and percentage of time equals how much money. Josh calls me one of his five soulmates. He and I spent many nights just sitting around and philosophizing, shooting the shit; this was usually like three in the morning, and we

were the only ones left in the giant ten-thousand-square-foot empty
loft, and I'd see him clicking and figuring out the take, and he'd say
"Good—we did nine thousand dollars today," rocking with that little
autistic body rock that he has.

**JOSH HARRIS**   Prodigy was looking for a new CEO, and when
they made the offer to Ed Bennett, the headhunter asked me if I would
talk to him. They trusted that I knew the racket. So Ed asked me if he
should take the job, and I said definitely. It was him versus some IBM
guy, and I'm worried about having him take away my contract, and
they always kept me on edge for that; it was a sick company. I mean
sick in the sense that it was literally not a healthy company. And so I'm
always worried that they're going to take away my contract, even while
I'm doing twenty-five percent of the traffic. Their view is, If he's doing
it, why should he be making all that money? As opposed to, Can we get
the guy to burn more hours. They didn't reward success. So it was Ed
versus some IBM guy, and Ed seemed like a much more interesting
choice. So Ed came along.

I could see that the world was changing. Prodigy was going to a flat-
rate scheme, something like that, where I'd be screwed. And also the
chat platform they were using really wasn't that smart, and the one we
were using was no big deal. So we decided to go build a new platform,
to build a project that was half owned by each of us. Which, of course,
being in the scene and knowing Ed and Betty [Rosserman, Bennett's
girlfriend at the time], I decided that it would be good to call the project
Project Betty. What the heck. I knew as long as they were together I'm in
good shape.

Then these new guys came in, and for whatever reason they bought
our half out—they bought our whole relationship out, and paid us a
lump sum of money. And that funded us for a period of time.

**ROBERT GALINSKY**   They started to realize that part of Josh's
genius was that he had his guys write the chat software for Prodigy,
and then he licensed it to them with some conditions: you can use your
software if you give us an area where we can do Pseudo chat next to
Prodigy chat, with our brand on it, and we make X a percentage on

numbers of people in the rooms and usage based on time. So Prodigy went for it, signed on for a year. Little did they know though that Pseudo chat was doing like eighty percent more traffic than Prodigy chat. Their own chat sucked. Because they had rooms like Connecticut and the Knitters' Room. Josh's rooms were Vampire Pub; the Neighborhood Bar; Domination and Submission. We were doing incredible traffic. They realized when their contract was coming up for renewal they weren't going to do it that way. They were embarrassed, and Josh is very arrogant. They didn't like it that Pseudo was beating them, and that Josh Harris was the guy behind it. We couldn't have our sugar daddy in Prodigy anymore.

**JOSH HARRIS** There was this woman who just started hanging around and staying at my house. She was there for like three months. She sort of stayed in my room. And I think we did it once, and I didn't really want to do it. It was like a mercy deal, to be perfectly frank.

She went crazy. I was running a fairly serious chat application on Prodigy, getting the software built, so I've got this great gig over there. And meanwhile I've got all hell breaking loose at home. From whatever party, I had a case of six half gallons of vodka. And I started noticing about one would be missing every two days.

And the last they found her, Jacques Tege, who I started Pseudo with—and also the animator that I made *Launder My Head* with, which I'm about to go into production on—he was downstairs at 600 Broadway, in his car, and she jumps in his car naked in the middle of winter. She'd just gotten out of a limousine filled with Japanese businessmen who I guess she was with all night. Because I'd finally kicked her out.

The night before, I'd been sitting in my room. She could see my light on, and I could hear her calling up to me, and I didn't go to the window. I just couldn't do it. I couldn't do it. It's like, I had to stop. It had to stop somewhere. And the next night she wound up in the car with the Japanese businessmen, and she wound up jumping out in subzero weather into Tege's car, and then running down the street naked with bloody feet. And then wound up in Bellevue for two or three weeks. And now she's fine, she has a kid and she's fine.

*What were the parties like?*

**ROBERT GALINSKY** They were great. There was one party that—it was the stupidest party. After it was over I realized how stupid we were. We could have been on *America's Most Stupid Tragic Videos*, hosted by Jack Palance, saying "What were these people thinking?"

We worked with a group called Fakeshop—Jeff Gomperts, really cool guy. And we lined the entire space—a ten-thousand-square-foot loft—with black plastic: ceiling and walls, every square inch, doors covered, tailor-made for the doors. We then opened up our fifth floor and had a two-floor party. And our tobacco sponsor was Moonlight cigarettes, and Skyy Vodka was the vodka sponsor. So we had cigarette girls walking around in push-up bras—among many other wild things, like people in crocodile costumes and people walking around with cameras on their heads. But the wild thing was that by midnight there was so much smoke in the space that literally you had to sit on the floor to breathe better.

So people were sitting down at the edges of the walls, just because three quarters of the room was filled with this haze. And they had given away something like twenty-two hundred cigarettes throughout the evening. That's when I realized that Jack Palance would have said, "They lined a building that's two hundred years old, that has one of the only wood foundations left among buildings its size—a landmark building, because it's all wood—with plastic on every wall, and gave away twenty-two hundred cigarettes, and overcrowded the place." There was something like four thousand people. It was a total death trap.

There were parties that were just purely outrageous, just for the sake of having a party. They were always strategically positioned in Josh's mind—and many people might not have known it—to court VCs or investors, to make position plays in the press. So there was always some ulterior motive, in terms of climbing the social ladder or the fiscal ladder, that some of us underlings didn't always understand. So some people would be angry that the PA sucked; we got the wrong PA; the video projection that I had wasn't right. Little did they know that it didn't matter in the big picture. The big picture was that two thousand people showed up, and Peter Gabriel got turned away at the door. It was worth the party just because Peter Gabriel got turned away at the door.

You'd have to pack your desk up, wrap it in plastic and push it against the wall, and now it was a serving table. You had to put your computer away; anything personal that you had would have to be locked down. Basically, when we had a party, if the party was Friday, by Tuesday you minimized your full-time job, just making sure the radio broadcast went out. But the next four days you were breaking the space down, metamorphosing the space, and concentrating and getting that party done. The party happened, and you spent the next three days digging yourself out from the party—literally digging yourself out.

There was one party we had—a Madonna party. Josh was fixated on having Madonna's baby. He wanted his sperm to be inseminated into her. So he was obsessed with how he could make noise on a level that would get under her skin so that she would consider this.

So we did a Madonna party. We auditioned Madonna acts, and we ultimately picked eight acts. We had some Japanese chicks doing karaoke to Madonna songs; we had this really serious actress who did this monologue from Madonna from the Bible; we had female impersonator Madonnas. And then we all had the opportunity to make some artwork, and Josh paid to blow it up on these incredibly brilliant six-by-four-foot giant prints, two thousand dollars each, and we had a gallery show.

**JESS ZAINO** | HOST, STARFREAKY | I was living in California, and I came to New York for a two-week vacation. Things weren't so good for me in California, and I needed to just chill out and be with my family. This was in December of 1997, and I was twenty years old.

One night I walked into Satellite Records and was like, "Where's a good place to go where I can go inside and be warm?"

And the guy there was like, "You got to check out Pseudo. It's this great place, this weird Internet company. They have music." So we went there, and they happened to be having a party for Levi's. I got totally drunk. I remember seeing young people all over the place, just doing their thing, and I remember thinking I had to be involved in whatever this craziness was. So I got totally drunk and I talked to Tony Asnes [then president of Pseudo].

I said, "What do you do?"

"I'm a janitor."

"Yeah, that's cool. I would lick the floors to work at this place."

So we started bullshitting, because I had these ideas, and I just started throwing them at this guy. And he was like, "This is great. Here's my card. You're a little drunk, I'm a little drunk. Why don't you call me tomorrow and we'll talk more?"

So I called him the next day and scheduled an interview, and he had me meet Robert Galinsky. I started immediately as their reception-ist, because their receptionist was awful and they wanted to get some-one new in there right away. I was like, "I'll do anything to work here."

So I started in '97. I had never worked in Silicon Alley, and I didn't know anyone who worked there. I hated the Internet before I got into it. I always thought it would be the end of human communication, and I was against it. But I started working there, and all of that changed. As the receptionist I was also Josh Harris's and Tony Asnes's assistant—their liaison to the outside world. I would answer the phone—"Hello, Pseudo!"—and make friends with anybody who called. And when peo-ple came in, they always knew who I was—I was the crazy girl on the other end of the line.

**ROBERT GALINSKY** As the network got defined, the parties started to get redefined. At the beginning there was only one radio show, with a bunch of creative people, so when we did the party it was a bunch of creative people doing the party. As we changed it became a radio network, with different shows in different genres. So in our com-pany you'd have the dudes who were totally into rock 'n' roll working on rock 'n' roll shows, and those who were totally into hip-hop work-ing on hip-hop.

My thing was, I was the executive producer of the network, and I was into the theater and the poetry and the fine art area of the site. And then you had the electronic music party. And as that got defined, it made its way into the parties. So, for example, at Roseland one room was folk rock. The kids—when I say kids, whatever—who did the rock 'n' roll shows for us programmed eight hours of singer/songwriter acts.

The Fakeshop guys did the largest bubble they ever did. They filled the entire Roseland with forty-four thousand square feet of plastic and inflated the room to a giant silver bubble. The whole room. And they had two of those roll-out, hydraulic lifts that go all the way up to the ceiling, and they had girls and guys doing seminaked dancing, way up high. It was really cool.

The floor was covered in plastic, so you were in this completely different environment. It was brilliant.

And then upstairs, in the back, I had programmed eight hours of spoken-word and music. I had guys from Groove Collective continuously playing music and pumping spoken-word artists through there, so upstairs there was this really highly politically and socially charged commentary going on.

And we bought bean bag chairs for that one, so people were just laying out and smoking reefer like nobody's business. And downstairs you had more people tripping out on acid in this giant silver room. It was amazing. It was amazing.

**JOHN EGAN PRESTON** | HOST, PARSETV | I was eighteen or nineteen years old, living with my parents, and I really didn't know what I wanted to do with myself. I found the Internet and my computer as an outlet. It really seemed like the only thing that was fun to do at the time. The problem was, it wasn't really paying the bills. And it wasn't really anything that, you know, my family could understand as something to do with your life.

My older sister was living in New York at the time; I was living in New Jersey. She knew I was playing around with my computer, and that I was also disenchanted with what I wanted to do with my future. She suggested Pseudo.com because she was friends with this person named Josh Harris, who apparently was the founder. And I think she had done stuff with Jupiter, or knew people who did. It was her belief, and the belief of a lot of other people in New York, that the Internet was going to revolutionize the way we do everything. Even at that time she was talking about broadband and online video.

She was saying, "This is something that you think is a hobby, or a way to kill time when you're bored. But the reality is, in the very near future this is going to be what the whole world is dependent on, and you should really hook up with these people and do something with them." And she also talked a lot about how Josh Harris and his company Pseudo was a real liberal atmosphere, and they were the kind of people I could work with and not feel the constraints or the stuck-up atmosphere you'd find in a normal office job.

She gave me the address, and I walked in there and asked for Josh Harris.

*What did you think of Josh when you first met him?*

**JESS ZAINO** I didn't know his story. It wasn't until a week into working that I hooked up with Uzi Fisher, one of the founders of Freq.net [a weekly electronic music dance party and show]. Uzi had brought me on to cohost Freq with him. So I was working as a receptionist, and then every Thursday from twelve to two I would cohost Freq.

Then, after a week of working there, somebody came by and said, "Why don't you come upstairs to Josh's place? We're always smoking weed up there." So we go up to Josh's house. He used to live on the sixth floor [of the Pseudo building], in the art department. I remember walking up to his house with maybe six other boys who worked at Pseudo, and we all just got really high and talked about work. I loved it. I felt like such a part of it.

I saw Josh was wacky, and I'm like, *I want to be friends with him.* But he's a hard person to get close to, just because of who he is. I think he just liked that I was always talking about celebrities, and I was always bullshitting and networking. So he would always give me props as his little girl, even though they were a little misogynistic. They were *so* about men there. It was a total boys' club at the beginning. The men were running it, and the women had very little say. They would have little outings where they would go smoke cigars and shoot things on the weekends.

**JOHN EGAN PRESTON** I wasn't able to get in touch with Josh Harris, but I ended up talking to one of the techies who worked there. And they were like, "Yeah, why are you looking for him? What's your story?" And I told him I was interested in working as an intern. Basically doing anything, as long as I could gain some experience by being there. That's all I wanted to do. And without even talking to Josh, without anything having to do with him, I ended up working with this guy who at the time was the network administrator. I would come in, you know, in the afternoons and do monkey work, which I was happy to do because it was just really cool to be working in a place where everyone was really laid back and young. I had never seen anything like that before.

**JESS ZAINO**  We were all these twenty-year-olds talking about how many shares we had. And we saw Josh making all this money. We all thought we were going to make money, and I begged for my shares. It was early on, and I was able to get a lot. I remember telling my parents, "By this time next year I'm going to be a millionaire." That was, like, two years ago.

**ROBERT GALINSKY**  I knew I would have to put my theatrical career on the side burner to make value on the shares I owned. I figured they'd be worth three million dollars, and that was at low- to mid-valuation, at ten bucks a share. It was always something imaginary in my back pocket, but I knew how I would spend it if that came through. And that was to make my own theater in Manhattan—small theater by night, media chop shop working there by day, have workshops for kids, and in the evening have great performances . . . it's still the dream.

Every six months Josh would say, "Just hold on—another six months." We'd go through iterations of people dropping off, and semisuicides, and Josh would come into the company meeting—at first it was twelve people in the company—"Just six more months, we'll have more money." And six months later there's twenty-five people in the company, and it was still "Just six more months." Six months later there's thirty-eight people in the company. After about two years, I started to realize, this is a long haul.

**JOHN EGAN PRESTON**  I was blown away. I thought it was really cool. Because at that age I was going through a period where I didn't feel like I could get into anything that was part of the real world. It was really refreshing, because everyone there seemed so liberated. They were young people; they could dress the way they wanted, talk the way they wanted. They could, you know, make their own hours, and do their own thing. But at the same time they're a productive entity as a whole. I mean, it gave me the impression that it was a place of business. This was a place of importance. But at the same time, everyone was really cool and having a lot of fun. Um, it sort of reminded me of high school.

**ROBERT GALINSKY**   In the first few months it was a clubhouse; it was great. All these stories about people smoking pot and doing coke and drinking on the job—that was a handful of people. I'd say snorting coke, very few people. Drinking on the job? A few people; the number goes up a little bit. Smoking marijuana? Could be half the company within the first year, on the job.

Marijuana was definitely a positive, recreational, creative enhancement to a lot of the work that we did. It wasn't a negative thing, it wasn't frowned upon, it wasn't something that was looked at as a bad thing, it wasn't discouraged other than just to keep the smell down in case any businesspeople showed up. But it got to the point that after ten months we had to say, No more smoking until six P.M. Ten to six—those are work hours, you don't smoke. If you do, you leave and come back. Six P.M., that's when you could smoke. At six P.M. the intercom would light up: "Code green! Code green!" We had language for it. We'd run down to the third floor and just smoke our brains out, and then go back up to work. And that was the thing—you'd work from ten to six, smoke some pot, and then go back to work until two A.M. It was part of what we were doing. It was great. It was so good. It was so wonderfully warm and accepting.

In that culture, people start showing up late. So Josh needed to figure out a way to get people to start showing up on time. One Monday, you got off the elevator on the sixth floor and there was a wooden desk and there's Josh with a video camera. From ten o'clock on—if it was 10:01, it didn't matter. "Tell us who you are and why you're late." It was so humiliating, because some people would say, "It's only 10:01."

"I don't care. Who are you? What do you do here? And why are you late?"

Some kids would walk in at ten-forty, and anyone after ten-thirty, he'd tell to go home. He wouldn't dock their pay, but he would tell them to go home and come back tomorrow. It was befuddling to people, because they could go home and were still getting paid— *What kind of punishment is this?*

There was a group called Half-Baked Productions that would sup-ply us with ounces of marijuana. If there was a party it would arrive two days early, and we would test it out before the audiences showed up—we didn't want to disappoint them. Josh and I would make sure we had a couple of ounces there, just like you would have a keg or a bar. We'd have a nice community of smoking. And of course it would still be there for a few days. But it wasn't like we had an in-house bong fixture.

**JOSH HARRIS** I smoke pot, I will admit to that. Now I'm kind of a teetotaler; I probably only get stoned once a month. But back then, maybe once a week or more. What can I say? I get bored easy.

People smoked pot at parties and that was fine by me. The other drugs were not [available], and even hard alcohol I never liked serving. Half Baked was a regular at the parties, and they did two things that were cool. One was, they made incredible bongs. They were bongs that were social, with six pipes so everyone could use them at the same time. I particularly liked that they ran it as an "installation," and we would put it in a room, close it off, make it discreet. The second thing was that they had a pot priest, so that while you were doing the pot they did a ceremony. People liked it; they appreciated the ceremonial factor.

**JESS ZAINO** There were a lot of people either dating or fucking. You always knew that someone was fucking somebody. My last two long-term relationships came from Pseudo—it was very incestuous. Weird things would happen at Pseudo. At weird hours of the night you would find people in corners doing weird stuff.

**ROBERT GALINSKY** I had sex in the middle of a party once up in the back, with my wife, who was not my wife at the time. That's the quality of the parties: You had strangers fooling around, but you also had people who were deeply connected to each other, taking chances, having a good time.

**JOHN EGAN PRESTON**   There was a lot of E, a lot of coke. Some of those rock 'n' roll people smoked a lot of dope. It was really anything goes, and what was so crazy about the parties was that they really brought it all out into the light.

**JOHN YOUNG**   The Pseudo parties were really good. They had a party for the release of one of the Quake or Doom games, and there were all these projections of the game on the walls, and a hundred people sitting at consoles playing the game, and it was such a visual overload. There were all these nooks and crannies, and someone pulled me down the stairway saying, "Hey, look, here's an Avid studio." All this free-wheeling abundance. And I was just thinking, "Whose trust fund is this?"

**CLAY SHIRKY**   Pseudo was always the tree fort of the Internet. It was a fun place to go if you didn't have anything particular to do. And in a way, the weird thing about Pseudo was that there were a lot of people there trying to do really good work, but often in the face of the existing culture. It was just sort of this weird grab bag: You'd go into the studio; there's real TV cameras and they're doing real streams. And they never say how many people are watching the streams.

The only thing good I ever got out of talking on a Pseudo show was meeting my fellow talk-show invitees. I don't think anyone ever listened in. But it was a weird place. It was our version of [Warhol's] Factory. And when it closed, and when Word closed, and a lot of these places closed that had been stringing along—"Well, it's the Internet, so it's got to be profitable some day"—that to me was the end of an era. Not that Silicon Alley was set up to support cultural institutions, but the way we do business is different from the way other people do business.

**ROBERT GALINSKY**   One of Josh's things was, We're not a rental house. We probably would have been a more successful business if we had taken our machine and rented it out more, because we were doing all our own design, all our own production. If we had taken some freelance gigs we would have rocked—we would have made some dough. But his thing was always *Stay pure—do our own work*, which is a great

ethic. And *Stay online.* People were always getting distracted and saying "Let's go do a movie," or "Let's go do TV," or "Why aren't we doing
more radio?" And he would always say, "Stay online. Just stay here,
stay pure. Know your audience; know your product; keep burning it in,
making those mistakes, correcting those mistakes. We're ahead of
everybody else." And that was great.

*What was the wildest party at Pseudo?*

**JESS ZAINO**   The party Uzi Fisher threw for New Year's '98. Uzi
decided to throw a party in the space, and asked Josh, who was away
for the weekend, if he could use the space. And he tells Josh it's only
going to be a couple people. So he promotes the fuck out of it, and
charges people.

There were people down the block trying to get in. They had no
security. Some kid had his eye stabbed out. There were fights that broke
out. The police and the fire department came in.

That was the end of Uzi forever. They were like, "You are *so* fired."
Josh and Tony fired him, and he was like Josh's illegitimate son. He was
just such a wild card, and they couldn't be sure that he wouldn't do it
again, and what if a million-dollar lawsuit came in from the kids who
came in there that night?

I remember being drunk and opening my eyes to see five hundred
people jumping and punching one another, and thinking to myself,
"Oh my God, what is this?" That was the craziest party I had ever been
to there—in terms of, just, craziness.

But there were many times when they had parties on the third floor.
When we first bought the third floor it was just open space, and they
would open it up for wine tasting parties and movie premieres, and a
lot of cool stuff happened down there.

**ROBERT GALINSKY**   Dennis Adamo was the first executive producer of the company. When I came in he and I respected and loved
each other, but there was tension because I was executive producer of
the radio show and he was executive producer of the company. So I
had control of how the show was done, taking in his input, but ulti-

mately they were my choices. And he was great for Josh because he would scream and yell at Josh, up to the point of being disrespectful, to get Josh to understand not to give it away. He was giving a lot of it away.

*How was he giving it away?*

**ROBERT GALINSKY**  Business choices. Who to invite, who not to invite. What kind of percentage and cuts to make. Dennis was very shrewd; he was a backbone, very objective. He wasn't very connected to anybody, or any of the companies. It was just about make money, snort coke and make money, snort coke and make money. Dennis was a real angry Russian-American dude—good Russian temper. Insanely deep partier. Coke and smoke and booze. I love Dennis and respect him and he can say anything about me.

About eight months into it, or maybe a year and a half, we noticed that the quality of Dennis's work was going way down. What made him cool about going out and snorting coke with business people, to keep us in the loop, was now becoming a negative. So we had an intervention. We pulled him into the space. There were eighteen of us—it was a big family back then—so we ambushed him and said, "You can't work here anymore if you keep doing this." One of our guys was in AA and suggested Hazelden, in Minnesota. We said, "The only way you can keep your job is if you go to Hazelden, and we'll pay for it." It was thirty thousand dollars.

It's the kind of thing that people don't know about Josh. Half the company was just like, "Cut him loose—he's a coke-freak loser now!" and half said, "No, he's a good person just going through some shit." And Josh said, "Look, I'll foot the bill. I'll pay the thirty Gs. You go to Hazelden for thirty-five days. You get your shit together and come back." Dennis had no choice—he knew what was happening—and so he went. He went out and came back a new man.

But while he was gone—thirty-five days in Internet life is about six months—the company was already rolling in a different gear, and personnel had already changed, and there was really no way for him to fit back in. And that's when he ended up leaving the company.

**JOSH HARRIS**  My epiphany in the last two years—the main moment for me—was about a year ago, maybe a year and a half. I was raising money for Pseudo, and I was stuck.

I learned fund-raising from Andrew Weinrich of sixdegrees, whose cousin is one of the angel investors in Pseudo, and also an investor in Jupiter, where he made a lot of money. So I'd gone to see Andrew, sat in front of him. He'd just raised ten million dollars. I was going nowhere, so I said, "How'd you do it, Andrew?" Well—he waited till he'd gotten his money to spill the beans—but he said, "Look, I just got on the phone, every night, every day. I'd get in at ten in the morning, I wouldn't leave until seven-thirty at night, and called these guys until I beat them senseless."

Oh, no problem. I figured I could do what he did in three months. Seven months later I'm walking on Great Jones [Street] heading toward the Bowery Bar, I'm waiting to cross the street, and I'm pretty screwed up. I'm like, *Damn,* and I can't snap out of it, and I'm beating myself up. And I had this little light go off, like, "Wait a minute here. I'm getting pissed off, but you know what's amazing? I'm in an amazing situation. If I can't raise the money, if I can't get this done, then I'm an asshole, and I can't blame anyone else. There's no—this is a pure situation."

And I realized, "That's why I came to New York." I came here to be in that situation. And all of a sudden, it's like I relaxed. Sure enough, then money sort of happened, once I hit that point, because they could feel it. I wasn't pressed. It was like, "Fuck it." If I fuck it up, at least I can look in the mirror and know it was me—it was pure me.

**JESS ZAINO**  Pseudo was an open space for relative freedom and expression. That all started to change when they started to hire more people on, and the business people were like, "We're going to make this good." And we were like, "Fuck you, it's good now." They were like, "We need to comply with the laws and not have kids up here anymore. There won't be any more drinking and smoking." It turned wacky, because we were all used to sitting at our desks and smoking joints, and that changed drastically and everyone felt it. They would send out these weird emails like, "You're not allowed to smoke in the building anymore."

**JOHN EGAN PRESTON**   For me—and this is going to sound like a bunch of bullshit—money never really played into the equation. The whole drama for me was being a part of something that was ground-breaking, revolutionary.

My sister, when she first started telling me about Pseudo, talked a lot about broadband. A lot of what she was saying about broadband was coming from statistics and projections that were being made by Jupiter. And I took that into account, and thought, Aw, man, broadband is gonna come into every household in America. It's gonna be the new medium of entertainment, and this company is capitalizing that. Not only that, they're pioneering it. They're the first ones to have an online presence that caters to this type of entertainment. That's what I was really more hyped up about. And as the company was growing I really began to feel like that was what was happening, that this must be what it was like to be part of one of the early television networks or radio stations. Money, and, you know, the other things associated with it, weren't what got me interested in it. Those weren't the things that made me think what we were doing was important.

**MARK STAHLMAN**   Josh is an interesting character, but he's not New York. He's Los Angeles. His training at the Annenberg School at UCLA is television training. So it was only natural, as Josh pursued his plans, that he would make the fundamental mistake of trying to put television on the Net. And no amount of attempts to rationalize that can really get around it. It just was a bad idea. Not too early, not too late, just a bad idea. Because it came from a different source, came from a different impulse to begin with.

In my first really close conversation with him many years ago, when Josh was telling me about his college background, he said, "We're all brainwashed; my job is to allow people to brainwash themselves." And that's what he set out to do with Pseudo.

**JOSH HARRIS**   We are really just a CPU and a hard drive. That's really what we are. Everything else is kind of a shell to guide that thing along. And we are being programmed, for better and for worse. So if you can put a layer of programming on there that says, Let's at least program ourselves with more consciousness about it and manage the electronic calories we consume, then it's better, maybe. Or it's not—we'll see. I'm not sure whether it's better or worse, but at least it should be something we look at. I think the business is programming people's lives. I direct reality.

**JESS ZAINO** I remember when Jeffrey Katzenberg and Brian Grazer of Pop.com [a failed DreamWorks project] came in to check Pseudo out, because they wanted to see what we were doing and how it looked. And I remember loving the fact that there was graffiti in the hallways, and this raw open space with wooden floors. Everyone's belongings would be there, and crazy paintings. Art everywhere—you would find some six-foot naked doll in the corner. And the elevator was a piece of shit—the slowest elevator in all of New York. They tried to cover up the graffiti for a while, because the investors didn't like it, but that was just who we were. Every night some kid would come in and just tag the place up. I can't tell you how many times I would come in Friday morning and there would be someone scrubbing down the elevator with bleach because some kid had tagged it up. It was such an amazing space—and it was in the most amazing place in all of the city, in the most amazing building.

**JOHN EGAN PRESTON** I went to a lot of other dot-com parties, especially at that time. And while the common theme is excess, the standard industry dot-com party was about excess in the sense of, *Look how fabulous we can be*. Whereas like a Pseudo party was about *Look how degraded and . . . depraved we can be. Look what a mockery we can make of ourselves*.

**JESS ZAINO** Josh is obsessed with old shows like *Gilligan's Island* and *Three's Company*, and he just loved the name Lovey. That was an alter ego of his that started on the *Pseudo Romance* show. One day he just came in with all this clown makeup on, and everyone was like, What the fuck is this? I was the first one in the studio with him, and we were locked up in the studio all alone. He was going "Boing! Boing!" I would ask, "Lovey, tell us what you are about." And this went on for twenty minutes. Me talking to him while he made *boing* noises. It was so freakish. That built into the whole *New York* magazine article, with that disgusting picture of him with drool coming out of his mouth. *That* is a genius visionary for you, right there.

**JOSH HARRIS**  The odd thing about Lovey—I'm dressing up in these weird outfits, and I'm CEO of this company that's funded by Intel and Tribune and all that. But I've learned the number one rule of New York very early on, which I impart to anyone I work with, and that is, Don't fuck with the money. So whatever I do, I never fuck with the money, and somehow they always know that's my prime directive. Even if I don't need to get money from them in the future, I protect their money, always.

**MARK STAHLMAN**  I don't think Josh really understands that people view him as a clown. Allowing himself to be portrayed as Lovey in that article is truly amazing. When you think about what people will do to themselves, it's really quite bizarre. But I sort of concluded that we would need a lot of this, to prepare New York for its longer-term role.

**JESS ZAINO**  All these new people came on, and we got a lot of money. We got fifteen million dollars' worth of financing. People were being hired by the boatload. This was in '98, early '99. Pseudo was lost forever after that, and we all felt it, and we were all sad about it but we plugged on.

**GENE DEROSE**  I really did always believe there was value there. Ironically, Josh needed to leave some version of his concrete role there, and effect a transition somewhat like what happened at Jupiter, but the handoff was bungled. And as a result it turned out that Josh leaving *was* the beginning of the end there. And that's a long . . . that's two years ago, maybe even a little bit more. But I really did believe, even up until a year ago, that there was a possibility it would go on.

**JESS ZAINO**  When we got the third floor for a reception area, I asked Tony to move me to something where I could have some creative input. He said, "Jess, you are one of the more colorful people here. I'll give you something to do." So I became associate producer of The Luscious Channel, which was women's lifestyles and sex and stuff. And there was this show called *Pseudo Romance*, and on that show I did a one-minute segment where I would talk about the lives of celebrities, and that eventually grew into a five-minute segment. And that eventually grew into a full-time show. I never wanted their format of a fifty-minute or half-hour show. *StarFreaky* started out as a show called *In the Bedroom*, and my first guest was [TV actor] Steven Weber. Then I just started to do *StarFreaky* full-time. It became the sleeper hit of Pseudo. No one thought it would be a big hit. I would get all these celebrities on the show and do all this weird irreverent shit, and people really dug it. And eventually I became the face of what they always put forth.

We were all superstars. Josh coined the Internet pop star, and he wanted to make us all famous. And we just became that. We just walked around like we fucking owned the Internet world.

*Josh Harris tried to bring in top executive talent to replace him at the CEO position, just as he had successfully done with Gene DeRose at Jupiter. But by early 1999, there was a tremendous talent shortage in the Internet industry, and the market for experienced top executives was prohibitive. Harris first tapped Larry Lux, head of National Geographic Interactive, to take over as CEO, but Lux never really caught on with the staff and resigned within nine months of his appointment.*

*By late 1999, Harris was already phasing himself out of the company, spending most of his time on his plans for Quiet and other projects.*

*Tony Asnes, longtime president of the company, was named acting CEO until a new appointment could be made. David Bohrman, a cable executive from CNNfn, became the new CEO in January. A small victory occurred when Pseudo won the right to Webcast for the first time ever the national political conventions in early 2000, and the Smithsonian Institution collected one of Pseudo's cameras for its permanent collection.*

*After wrapping Quiet, Harris took a trip to Africa with friends, and returned in the summer of 2000 to begin his next project, WeLiveInPublic.com, which entailed wiring his SoHo loft with video cameras and microphones, and putting his life, and the lives of his girlfriend Tanya Corrin and his two cats, on public display over the Internet.*

*Bohrman raised a fourteen million-dollar round from French media and fashion company LVMH. But the burn rate remained sky-high at two million dollars per month, and the April 2000 market crash ruined Pseudo's attempt to raise an even larger round over the summer.*

*In June, Bohrman announced layoff of fifty-eight workers, which many analysts thought to be a good move, and then dropped the format of programs on channels in favor of a single live discussion run by the Pseudo talent that would air online without interruption. The new nonstop format never succeeded in attracting a critical base of users, and last-minute attempts to raise capital all failed.*

*With Pseudo collapsing under the weight of its production costs, and failing to IPO or sell, Harris himself loaned a million dollars to the struggling company. The collateral was the entirety of Pseudo's recorded programming—"the best record of the emergence of the digital revolution," as Harris later said.*

*On September 18, 2000, Bohrman fired all one hundred and seventy-five remaining employees, and soon after declared chapter-eleven bankruptcy for Pseudo. For all its contributions to Alley culture, Pseudo had failed to live out the first year of the twenty-first century.*

**GENE DEROSE**  When Bohrman came in, his decision was The Wrong Decision to go with, by definition. Capital *T,* capital *W,* capital *D.* He decided to make it a TV network model, with macrotopics, and I think Pseudo could even be alive today if they still had forty-five or more different micro topics—*and* a very radical approach to those. And clearly sane business management was needed, and it was never there. I really do believe about advertising and about content and creative and the niche-content side of it, but the typical pathology of these companies—and Pseudo was the most guilty of it—was to do the wrong thing. They would do step 2, step 3, and step 1—in that order.

There are three basic elements to what you have to do in this kind of business. You have to get the content and the product model down: It should have a little bit of technology, a little bit of creativity, a little bit of uniquely using the medium. There has to be some essence to what you do, besides, say, artistic racy downtown New York, forty topics. That's not enough. You have to get something down that's compelling. Category two is, you drive traffic, and you build an audience, or you prove that you can build an audience. You don't have to spend a lot of money, necessarily, to do that. Scott Kurnit has done an incredible job with About.com without spending a lot of money. And then category three is, you sell advertising on top of it.

Pseudo did what a lot of these other companies do: Go out and spend a lot of money, try to hire an ad sales guy, before they even figure out step one. They're simultaneously marketing, and trying to sell ads, when they don't even have an audience. It never works; you never get out of the gates. That was always the problem.

**ROBERT GALINSKY**   Josh slowly started moving out of the company, spiritually. He felt it had its legs, and good people in executive management positions. He knew there were some holes, but felt that those people could take care of them. That was part of his plan, to ease his way out. But unfortunately a couple of those key management hires didn't work out, didn't fulfill.

**JOSH HARRIS**   When I leave, I leave. I don't run the company. It had its time. It's the first Internet TV network. It worked while I was there; I left it financially in good shape. Could've maybe had a better business team, in retrospect. But what are you going to do? I did what I could. I left it healthy. I left a decent model. Might have been a little ahead of its time. That's really what we're getting down to. But I don't know. I would always think, of course, that if I was still there, and that's all I was doing, it would be in good shape. But then again, if I was doing that, I couldn't have done Quiet; I couldn't have done the stuff I'm doing now.

**JASON CHERVOKAS**   I think the right-minded people in Silicon Alley always thought there was something incredibly ridiculous and impossible to understand about what was going on at Pseudo. The thing always seemed loony, because (a) it wasn't a business, obviously—it was never run as a business, it was never built to be a business; (b) the notion of creating TV for an alternative distribution platform was ridiculous. We already had perfectly good TV. Why did we need something else? Down to its most fundamental assumptions, the whole thing made no sense. I think people in New York were so caught up in the sense of creativity and revolution, and the fun of the party scene, that there were a lot of people who didn't step back. Though, actually, people did: Pseudo didn't raise real money for years.

And looking back I think we should have taken it as the sign that the end was coming when Pseudo raised as much money as it did from what looked like sober investors. I remember thinking at the time, Gee, they got this money from Tribune and Prospect Street? These guys ought to know better. What do they see in this? And I spoke to some of those guys and asked them what they saw in it, and they were very much enamored of the fact that broadband was beginning to become a reality to a very small collegiate test market, and they thought, We'll learn some lessons here. But I think it was really a sign that a sort of investment mania had finally driven all logic out of the way we invested. I wish I had realized it at the time.

*What were the days like directly leading up to the close?*

**JESS ZAINO**  A week prior to it I was told by the head production people that I shouldn't worry about my show—that I didn't have to do it. Not to worry about going into production, because it was really crazy right now. I was totally in denial about it. I kept on working and doing my thing. I wanted to keep doing *StarFreaky* for as long as I could. Then it was on DotComFailures.com that we were going to close that Friday, and people kept saying it was just a rumor. But in fact we did close on Friday, because Friday was the last day that we got paid for.

The higher-ups were never around—they never talked, never sent any word out. No one knew what was going on. People knew in the upper ranks, but no one else really knew what was going on. The management was in a separate building; you never saw them.

Then on Monday, when everybody came in, there was an email saying we were all going to meet at eleven. Then David came in and said, "This is it—this is the end. We are going to give it another two hours to see if we can salvage some money, and if we don't get that investment in the next two hours Pseudo is going to close." He was sort of giving us the runaround, and I raised my hand and asked, "David, are we getting paid? That's the bottom line here—are you going to fucking pay us?"

And he made some snide remark at me and said, "No. You got your last paychecks on Friday." And everyone was like, Oh my God. And people broke down. People were crying. There was mass hysteria.

We were told to come back in two hours, so everyone went across the street to the Puck Fair. The Puck Fair became our drinking hole.

Everyone got wasted at the Puck Fair, and came back two hours later just to be told that Pseudo had closed and that that was it. The HR people came in to tell us what to do with our health insurance and our gym memberships and our stocks. "It's all gone. You have a week to get all this shit out." We were treated with such disrespect. I had a contract I'd worked really hard to get, where I had lots of severance, and I didn't get anything. Once they claimed bankruptcy, no one got anything.

**MARK STAHLMAN**   The collapse of Pseudo is probably the single most important watershed in all of this—along with the way people are attempting to explain why and how it happened.

Because it demonstrates sort of the problem with Silicon Alley. Has this been a bunch of people who want to become rich and famous because their parents are rich and famous, or because they were brought up believing it was their destiny to become rich and famous? How can you tell if that's what was going on? I think the easiest way would be how people treat the media. To the extent that people want to be on television, to be on the covers of magazines—that's one attitude. To the extent that people actually ask the question, What do you mean by new media? that's another attitude. And both those things have been very prevalent in the community—which hasn't *been* a community. There's no such thing as a Silicon Alley community. There's been a number of disjointed groups, and I think a certain amount of distrust has been embedded within the situation from the very beginning.

**JESS ZAINO**   The last day there were these police everywhere. They had these plainclothes policemen on the roof, outside, and they were escorting us out of the building. We were not allowed in the building for two days. It was locked down.

*They were worried that you were going to steal equipment?*

**JESS ZAINO**   Yeah. I think that that went back to a couple months before, when sixty-eight people had been fired and tons of shit got taken. Which I didn't know, until David actually told me that. There were police everywhere, and they were checking all the boxes that went out. We thought, Fuck you—we've been working here for years, and our lives are in these boxes, and you're going to open them up and go through them for the sake of finding a disk you don't want us to take? That was really rough.

**ROBERT GALINSKY** The night after it happened I went over to Josh's place and got spontaneously drunk and stayed up until six A.M., literally laughing and crying over the past, and what could have happened, and what we thought would have happened. It was definitely a purging. I talk to him every day. We have the same kinds of feelings. It's like somebody died. It's extremely tragic and sad, and completely unexpected. We're going through the phases of denial and anger, trying to not blame anybody. But then at the same time it's extremely liberating. It would have been far more liberating to have gone public and become a millionaire. But it's finally over—this pending possible success or massive failure is over—and we can get on with our lives now.

*If Broadcast.com was acquired for more than $5 billion, couldn't Pseudo also have sold out for billions of dollars?*

**SETH GOLDSTEIN** No, they had no technology discipline. And the Broadcast.com sale was a joke, too. All this shit is bogus. It's empty. There's nothing there. You can't point to companies and say, That was real; that wasn't. Page views, I guess, are real. Are banners real? Because someone serves a billion banners, does that mean they're a real company? Because nobody clicks on banners.

**ROBERT GALINSKY** One of the things that bothered me the most about the demise of Pseudo: Most of the reporting has been about how the business failed, and the market failed the business, and the managers sucked, and the model wasn't ready. But nobody has lamented or talked about the cultural loss, and that space, and the people that worked through that space. Nobody has. I've never heard anyone talk about how it was a true New York institution. So many people have passed through those doors, from the most intensely pure dedicated raver-style kids—twelve-year-olds to twenty-eight-year-olds, to the pure hip-hop kids who came out of Brooklyn. Every Wednesday hundreds and hundreds of kids coming through and looking for something new and hopeful and positive. Raw people from all over the boroughs, not only getting their particular genre fix but much else besides. That loss nobody has focused on, and that to me is the most critical loss.

And you would have people coming in from Europe and Japan and Africa, trying to cover the place and freaking out because it was authentic art—human life creating this organic culture.

**JESS ZAINO**  Josh gave us the opportunity to do what ever we wanted to do and materialize it. I didn't graduate. I was some stupid fucked-up kid on the streets when I walked into Pseudo; three years later my face is on the cover of magazines and I'm being quoted as being part of hip Pseudo. How the fuck did that happen? It's because Josh gave us that opportunity to do whatever we wanted, and we were allowed to run rampant with who we were. It was so much fun. And when it closed I was hysterical. Did you see my tattoo? It's the Pseudo logo.

**JOSH HARRIS**  Sometimes I wake up and feel that I let people down. And sometimes I say that's the price of tea in China; I gave a lot of people a lot of good years. Depending on where the moon is, I don't know.

I don't know why I do anything. It's like brain waves come in from the universe or something, but it just seems obvious to do it.

In the last ten years I've probably been to Bellevue thirty or forty times—visiting, not as a resident—everything from drug overdoses to suicides to people just checking themselves in. Just crazy things. And I feel like I've put a few people in—I'm sure I put five or ten in. I can quantifiably certify that I've certified them. So it's like, If I go there myself I'd go crazy. And I'm always on the edge of going crazy, so it's a total vigilance not to go there permanently.

I think I'll be bigger actually [than Andy Warhol]. Bigger or smaller, it's harder to say, because the way I see it, in the spectrum of time, he is my ad man—he made the ads for all the product I'm producing. I'm starting to realize that. And we'll see; it won't be that way until it is, but, if you think about it . . . there's that famous Warhol and Basquiat boxing poster; I'm going to have twenty of those for different fighters, and I'm actually going to have the fights. I'm actually making the product that he advertised.

Marilyn Monroe—maybe that's his signature piece, or Chairman Mao. I'm doing Gilligan. 'Cause Gilligan is really the body of work that changed the culture. Marilyn Monroe was the surface of that change, but it was really Gilligan, and the mores that he brought to a generation of culture, that really shaped their behavior and shaped who they are. Gilligan, to me, was a primal force in terms of other mediation than physical. That was my formative Wonder Bread, my formative years. Gilligan shaped me.

*With Pseudo behind him, Harris announced that for his newest venture he would be wiring his own house with cameras and living his life as an open book.*

**JOSH HARRIS** Investors still all like me, for some reason. Which is really hard to fathom, considering what I'm doing now. You know, I'm bugging my house—essentially Big Brotherizing my house. I'm doing a piece called "We Live in Public." Myself and Tanya Corrin, who lives with me back there—she does *TanyaTV* on Pseudo, so she's an Internet television personality—we are going to live in public. That's why all these holes are here, and that's why these guys are wiring and putting in forty cameras, and it's not normal. It's normal, but not average. I don't know what the fuck that is.

They're going to watch what we do all day long in this place—everywhere. Bedroom, bathroom, cat box, as many places as we thought would be interesting. It will go directly out to the Net; it will be stored on the Net for some period of time, and be recorded in my control room on tapes, and the tapes I'll sell as art. And if you think about it, work through the problem, getting the right people to come here under the right conditions makes the video that I make here more valuable—particularly if it's late at night and they've been drinking and they have to use the bathroom. I think I'm going to capture some of the world's greatest dumps, 'cause there's a camera facing upward in the toilet, inside the commode. Infrared.

What's happening in the world is that people don't want to watch other people on the cameras, they want the camera watching them. Because we've been programmed to idolize the things we see, and we've been also programmed to want to *be* those things. So if the camera turns around on us then we *are* those things, and it starts this weird vicious cycle where we want more people to watch us, and that's all we care about, instead of watching the other people. It's fascinating to me, and I *know* what will happen with me. It's part of the work so . . . It excites me more to watch a Yankee game or something like that, but I know with me I'll be curious to see what the dynamics are. It's curiosity.

**JESS ZAINO**   It's what he always wanted to do—good for him. I myself am freaked out by it, because I don't want my ass shown on some huge projection screen. But it's a good experience. He's a visionary.

*Early in 2001, WeLiveInPublic shut down and Harris made plans to leave New York City.*

# FALLING OFF
# THEGLOBE.COM

*T*he story of TheGlobe.com is a definitive dot-com saga, with all the promise of youth, greatness, and riches beyond imagination crushed by the brutality of the public markets and scorched by hubris and naïveté. Established by two Cornell students in 1995, TheGlobe was committed to the concept of community—based on the idea that aggregating users in chat rooms and on home pages would produce immeasurable revenue. Early-twentysomething founders Stephan Paternot and Todd Krizelman watched as the company IPOd in November of 1998—one of the first companies to IPO successfully following a crisis in the Asian currency markets that shut down technology IPOs for several months and caused a steep drop in the NASDAQ. With a spectacular first-day rise from nine dollars to over ninety dollars per share, TheGlobe set a record as the fastest gainer in history. The founders were suddenly worth over one hundred million dollars each on paper, and they quickly hit the press circuit to show off their success on magazine covers and in talk shows. But things turned sour for TheGlobe, especially after a secondary offering filed in 1999. The stock soon plummeted from its height; by the end of 2000, it had fallen below $1 per share. A company once valued at nearly four billion dol-

*lars was now worth fifteen million. And the stakes of its founders dwindled away.*

*Perhaps no one company in the Alley inspires as much vitriol among its peers as TheGlobe. Mention its name, or those of its founders, and you hear a litany of complaints: "Internet mania," "IPO craze," "overvalued," "inexperienced," "naïve"—the list goes on. Most Alley entrepreneurs blame TheGlobe for inspiring Internet mania, then poisoning the well for all dot-com stocks. But the young founders certainly lived through a remarkable experience.*

**STEPHAN PATERNOT** | COFOUNDER, THEGLOBE.COM | One night in college, Todd and I found a chat room online where you'd see messages scrolling by. It was like, "Oh my God, this is revolutionary." And after about four hours of getting sucked into this thing—it was probably midnight at this point—Todd and I just looked at each other and said, This is it. This is the mother of all applications. This is going to be the thing that never existed without the Internet. And we just started from there.

**TODD KRIZELMAN**   It felt like the Dark Ages. We were in Ithaca, New York, finishing degrees at Cornell, which was probably one of the most difficult places—at least in the United States—to start a company and, more important, to raise money. At this time the Internet was not popular. This was all before the mania and excitement. It was not viewed as a positive thing by our families or our friends. In fact, if anything, both those groups reinforced that this was a fun project for us, but that we'd better finish our degrees—which we did because we were so afraid that if the company didn't work out we would have nothing. Our fondest memories are looking back on how seriously we took ourselves in the beginning.

**STEPH**   Our own naïveté is the only thing that kept us going.

**TODD**   We were so serious. We would take out these full page ads in the *Cornell Daily Sun* for like two hundred dollars, which was a lot of money out of the fifteen thousand we started with. We would say, "Looking for serious management. You must be a junior or senior in college." These ads would look like they were for a real company, so we would get hundreds of responses. The first time we did it was such a mess. Hundreds of people were lining up into the student union.

**STEPH**   Todd and I were dressed in jeans and T-shirts, sitting in a hall where there's some guy banging away on a piano.

**TODD**   It's not like we'd rented a room or something, we just sat down. And I remember running back and forth getting sandwiches in the cafeteria.

**STEPH**   "Hey, welcome! Have my sandwich."

**TODD**   People were dressed up in ties and dresses; they had their résumés . . .

**STEPH**   We wanted people who were just as naïve as us in the beginning, who were just like, A hundred hours a week? Sounds good to me!

**TODD**   And it's not like we led people on. We'd say, "If you're looking to work a hundred hours a week, get paid absolutely nothing . . . this is absolutely for you. You're gonna love it. We're going to take over the world."

**STEPH**   We moved out of Todd's dorm room because we needed a professional office, and into a windowless storage room hidden away in the computer science building. There were all these boxes filled with old test scores and exams—which was great for Todd and me when exams came up. We would rummage through there and try to see if there were answers to the exams we were about to have. Todd and I erected a wall of old boxes so that we could have our executive area, which was just a disaster waiting to happen—the boxes just tipping over would crush us.

**TODD**   The entire room, I think it was a little smaller than this room. It was maybe seventy-five percent of this.

**STEPH**   We put up a wall right here, and I was like, "Ahem, yes. [seriously] How's it going over there? You can knock to come in. Sure."

It was pathetic. It was no bigger than a third of this room. Or a quarter of this room.

**TODD**   Oh no, it was bigger than that. It was at least half this room. But it was small. And there were no windows. No pretty track lighting. It was just . . .

**STEPH**   I think, Todd, you were smaller, so to you the room looked bigger. It was a small room.

**TODD**   Oh God, what a disaster that was.

**STEPH**   Those were the good old days.

**TODD**   And we didn't have enough computers, so people would have to share sometimes.

**STEPH**   Sometimes we'd take servers down so that you could use a computer to do something. Which was really ridiculous.

**TODD**   All of us loved it.

**STEPH**   When we started our business, Todd and I were such Mac addicts that we decided to build the entire business on Macintoshes—which is what we did. Of course, if you know anything about running high-end technologies, you do not run them off of Macs. You run off of Spark stations, off of Unix and all that top-level stuff.

**TODD**   We did it for two reasons. One, we loved Macs; two, they gave us free computers. We were in all their marketing.

**STEPH**   At that point our site was becoming one of the more trafficked sites—you know, back then when we had like fifty thousand users. I think at some point we officially became the most trafficked site in the world that was running on Macintosh servers. Which, if you were a real tech guy, you would've laughed at: "What a bunch of idiots." But at Apple they were like, "This is great!"

**TODD**  I don't know if we have those pamphlets anymore, but they were like, "Target, Valvoline, and TheGlobe all run on Macintosh." I remember thinking, This is impossible. Apple's going to go out of business. We're the best? There's, like, ten of us in the company. This can't be right.

But it was a fun time. None of us thought that it was dire straits. It was the most exciting time. It was a different environment. Steph and I constantly talked about the fact that we were on the edge of something no one else had seen yet. When we were growing up, then when we first started college, we had talked about how the computer revolution was something that had happened and we had missed it. I grew up in the [San Francisco] Bay area, and I felt it every day. It was something we talked about almost every day, because you read about Apple, and you read about Intel and Microsoft, and these were exciting times. Out there we went through the recession as though nothing happened. As kids of that generation, you felt as if you'd just missed out on the next industrial revolution. So when the Internet came, it felt like this would be our chance.

**STEPH**  I think also there was another built-in fear, which was that, as Todd mentioned, coming out of a recession there was a serious mentality across the college that we just wanted to be able to get any type of job. You wouldn't even have a choice; you were desperately looking for one. I remember early on, when I was doing computer science and Todd was doing bio, that the best jobs you could get as a computer scientist were either information technology at Merrill Lynch or information technology at Goldman Sachs. I could be one guy out of fifty thousand who would be their database guy, and I'd be paid fifty-three thousand dollars a year. That was what was pulling the other way. I was like, My God, that's the big bucks. And Todd and I at the same time felt like, God, how sucky is *that* gonna be? We were desperately looking to do *anything* else. Anything that would inspire us. And then we suddenly felt like we were in the beginning of something. We didn't want to abandon our degrees, but at the same time it was like, My God, if this thing could just take off we'd be right at the beginning of it.

It's actually sort of funny. We talk about what community was—but the term "community" didn't even exist in the Internet. It was just chat.

We wanted a chat room. We wanted to get people to interact. You know, what we eventually realized as we patched our site together was that, as you build your company and you need to approach investors, you're going to need a buzzword. The investors need a thing to hang their hat on, so they can say, "I'm getting into that sector. I'm getting into biotech, I'm getting into the Internet, I'm getting into 'community,' I'm getting into e-commerce."

**TODD** Another thing that was interesting about that time for us was the type of people you recruited. They were extraordinarily passionate about product. I mean, it wasn't that no one realized they had to run a business, but the people who got involved in this were just very much into the Internet. Our director of technology, Vance Huntley, who is still with the company and is our head of technology now, is a great example. He was the head falafel chef at . . .

**STEPH** . . . a local eatery. And we had to try to convince him to stay on board full-time, but he just said, "Guys, as long as I can only make"— I think we were paying him only six dollars an hour; he had to do his other job because he was paying his own way through college. The guy was working ninety-, hundred-hour weeks, to the point where he looked like a ghost. But those guys, you know, they were the early geeks, early pioneers. The guys who really had nothing better to do, because it's not like they were in party heaven, and Cornell's sort of lost in the middle of nowhere, so they had every excuse to try to get into something. I think a lot of these early guys who came in that way were programming just for the love of it. They were doing it anyway, so they thought, Hey, if I can make five, six bucks an hour for what I already love to do, great!

The stories of paying people by giving them free pizza are accurate. It was all Domino's, for the record. We did offer everyone options, but many people told us they didn't want the options—many people. You can't eat with options, they said.

**TODD** We would have people who would say, "I would rather have no options and an extra dollar in my paycheck." Even Steph and I used to joke that it was like Monopoly money, because no one even knew what an exit strategy was.

**STEPH**  You know how all those cars around the world that have Firestone tires have been criticized for their Firestone tires—as if every part of the car was still in testing when they shipped the final product? I mean, that's how bad it is in the Internet world. The wheels come flying off; the steering wheel is square; they haven't invented a round one yet. Meanwhile, you're just trying to come up with your little niche, trying to tell people how to spell "Internet," trying to tell people the difference between a browser and bandwidth. It was insane.

After the search engines went public and Netscape went public, it was the first time some investors started trying to speak the language: "Okay, talk to me. Bandwidth? Java? Do you guys have Java? You have to have lots of Java, because that's what's going to make it all work."

In the early days, nothing was perfectly clear. We didn't just go after community. We had to develop our own tools. So part of our business was deciding, Should we sell tools? Should we sell the applications? Where's it gonna go? So when you're trying to raise money in those conditions, and you've got five guys you need to pay, you make a decision. We couldn't afford to do software development and run a software company—and Oh, by the way, there were other established players that sold software . . . like Microsoft. So we said, Uh-oh, you know, are we really going to make an impact there? Because they could replicate this in a second, and then Game over! They have distribution, we don't.

We decided we wanted to be a media play [company] online. Now what the hell's a media play online? It meant that we would be all about aggregating audiences and trying to figure out how to sell advertising to them. And so we decided after our first year to drop the tools: Let's not go chasing after Java this or Java that, or Netscape plug-ins. There was a whole era when designing plug-ins was hot. And investors would tell us, You gotta do plug-ins, you gotta be Java. So instead we went the media direction.

**SUSAN BERKOWITZ**  DoubleClick was doing all the ad serving at the time, because nobody had an ad sales force. It was completely outsourced to DoubleClick. It's ironic, because the second we canceled the contract with DoubleClick to try to do inside sales, we actually stopped making money.

**STEPH**   It was hard for us, by the way, to raise money, because when we were flying back and forth to California, Todd and I were still in the middle of exams, and we were skipping class like there was no tomorrow. Our professors were like, "Guys, you're here for an education, you know?" We negotiated our way out of classes; we'd disappear for weeks on end, go out to California, stay at Todd's place. At this point I'm like his brother. Every single VC, at every point, was telling us, "No, you're too young, who the hell are you?" or "No, you're on the East Coast," and the last thing they'd always say was that we had to do tools. "If you do a cool Java chat we'll invest." We didn't want to do that; we wanted to be a media play.

We lucked out, though, because eventually we met David Horowitz, who was one of the founders and CEOs of MTV. That was a guy who understood media—and he was based on the East Coast. Already we had two things nailed, because VCs are so geographic in their investments; this guy was in the right part of the world for us. He was East Coast, understood media, and had a name. So we brought him in as an investor. Back in these days, again, large investments were like fifty thousand dollars. So we got fifty thousand dollars from him, which gave us another $X$ many months of life, but it allowed Todd and I to start playing this other game: "Hey, we got David Horowitz on board. Who's next?" And you start working your way up, trying to bring on great investors and board members.

**SUSAN BERKOWITZ**   David Horowitz, the guy who originally put money into *Spin*, was Bob Pittman's partner in founding MTV, and was the CFO of Warner Communications, was also an East Coast angel investor. He put money into this little company up in Ithaca.

David said to me, "I have these two guys, they're pretty young, not even finished with school yet, they have an office, and I just gave them half a million dollars [*sic*]. They say they have a hundred and fifty thousand users, and they do something they say is 'chat.' I'm not really sure what that is, but they say they'll have a million users by the end of the year."

The next thing I know, these two guys are standing in my living room. They were up in Ithaca, but they drove down one weekend, got me a computer, put it in my apartment, and that was the New York office of TheGlobe for a while. They would even come and sleep on my floor when they had meetings.

**TODD**  During that period, media plays became more interesting. In the first three years people had looked at us like we were crazy. People would put term sheets on the table that said, "Go do software. We'll give you a few million if you turn this into a software play." Ultimately it turned into something of value to say we had a media presence and that we were in New York.

**STEPH**  By that time, portals weren't even looked at as media plays; they were thought of as search engines—that is, as a tool. But of course they're media plays just like us. Everyone's in the advertising game, aggregating audience, but back then, when they started defining things, portals were defined as search engines, and we were defined as community.

The first three years were the toughest. We raised very little money. Our first year was a continuous fund-raising process. Todd and I just spent six months out of every year trying to raise money. The first year we raised a hundred thousand; the second year we managed to raise half a million. The third year we raised one million. After we finally graduated, the first decision we made was to get out of Ithaca. We needed to get down to a metropolis, where there's infrastructure and media.

That was probably the first major step to getting us into the big league. There *was* no big league at the time, of course, but it brought us to a point where we could be noticed. We got an article in the *New York Times* for having moved down to the city. That put us a little bit on the radar—and then we had the good fortune of meeting Michael Egan, who had just sold Alamo Rent-a-Car to Auto Nation and made his fortune.

Michael, who's a visionary and an entrepreneur himself, was looking for the next thing to get into, and he wanted to get into a media business. So you could very rapidly see how we made a connection. And right after we met Michael and he basically told us he absolutely wanted to invest, we then spent five months negotiating with him. That was our first taste of what it was really like to go through due diligence and get a real professional investor behind you. There was a war of attrition; these guys kept going and going and going until Todd and I were basically dead and tired, and eventually we raised the money— twenty million dollars. Which I think was defined at the time as the biggest investment ever by an individual in this space. So that put us on the map. People noticed TheGlobe.com for the first time.

**ANNA WHEATLEY**   I interviewed Todd and Steph the week after they got the twenty-million-dollar check, and our thing was to find companies before other people had written about them. We were one of the first to do an in-depth profile. And they were charming. Their office was on Eighteenth Street in Chelsea, and they were telling this whole story of meeting in college, and how they had given all this equity to get their friends to come down here. It was all about community, and wasn't this great and glorious. This is what I was thinking: "These are smart guys, and nobody fails in this business." However, I didn't understand that model—eighteen-to thirty-five-year-olds, and give them some "tools to build community," and what? But they were passionate, almost evangelical, in their belief in it. They had tried to surround themselves with really great advisors, which they had. But I thought, There is no way, at twenty-three, which is how old they were then . . . I couldn't even understand their articulation of the business, much less how they thought they were going to execute it, because doing a business is not creative in the way that writing is. Running a business means selling something, and that's the part that I kept missing. I thought, Community! Great. Bless their hearts, they believed it. But they were their own problem.

**FRED WILSON**   I think TheGlobe got to the game late in terms of community and building a community on the Internet, and they took advantage of the stock market to raise a lot of money, but I don't know that they ever built such a huge community that they had a chance to do something interesting.

**JERRY COLONNA**   That race was won by GeoCities, and if anyone else by Tripod. And if you look at what Bo [Peabody] is trying to do with Tripod, and what David Bohnett did with GeoCities—while they were similar services, they both had a unique feel about them. They were very different kinds of properties and were trying to achieve different kinds of ends. I never felt that TheGlobe, from the very beginning, had a unique value proposition, even in the community space. There was no center there. It was sort of a city without a downtown.

**STEPH**  For the companies that came later it was more obvious; there was money to be made—you could see who the winners and the losers were. You could see things much more clearly, and I think that if you went in with that mentality, you want to play it safer. Do I wish I'd cashed out more money and played it safer? Sure. But back then there was no notion of making money. Todd and I had never built businesses before, we'd never been burnt before, we'd never lost anything before. We had all the upside in the world. We had nothing; there was nothing to lose.

**TODD**  When you start with nothing it makes it very easy.

**SUSAN BERKOWITZ**  I remember getting to TheGlobe the first day. I had come from J. Walter Thompson, where I had this huge office, TV, a stereo, and leather chairs. And they took me on the first day to the art department, and I picked out a Lichtenstein and a Picasso.

I got to TheGlobe, and they handed me a box. It was my chair, and I had to build it. I'm like, Where's the art department? Where's my Picasso?

**MICHAEL KAWOCHKA** | FORMER PRODUCER, THEGLOBE | The woman who hired me at TheGlobe was a college dropout, about five years my junior—I was twenty-seven when I was hired. She did a little bit of Web stuff for Harpo Productions, Oprah Winfrey's company, before coming to TheGlobe. In the year or so she was there, she was promoted three or four times purely on the merit of her work and work ethic. A testament to meritocracy—this impressed me. It said to me, No politics or bullshit—kick ass and you'll get ahead. Besides, the co-CEOs were twenty-five, so gray hair, or other old-economy prerequisites, shouldn't be a factor in my career development.

And there were plenty of other old-economy corporate refugees like myself around the office, from companies like Andersen Consulting, Pfizer, Coca-Cola, and Saatchi & Saatchi. This said to me, Okay, I'm not the only one, so this must be a good place to make it in new media.

And I can't stress the word "casual" enough in terms of dress code. Friends of mine would drop by TheGlobe.com and turn green with envy. Shorts, sneakers, and T-shirts made it easier for me to get ready in the morning. Not having to spend ten minutes every morning trying to match my tie to my suspenders is indeed a perk. I sported a Knicks jersey all through the playoffs. Dry-cleaning bills were a distant memory. Sandals, tattoos, dyed hair were all nonissues in the workplace.

TheGlobe had the accoutrements: swank new office, casual dress code, ping pong, pool table, and weekly happy hours. This was all a culture shock coming from an uptight old-school corporate PR firm [Edelman Public Relations Worldwide]. I used to wear a suit four days a week and keep electronic timesheets of my billable hours. "Casual" Fridays meant khakis and no tie. God forbid you were ever late or left early. Paying your dues, kissing the right ass, and face time merely for face time's sake had more to do with your career advancement than what you actually produced. That's why I got out of that industry.

**CONNIE CONNORS**  I was in the governor's New Media Task Force, and they invited the Globe guys to come talk—this was before they'd gone public, when they were preparing to go public. People in this room were asking them about taxation issues. I think they appreciated this, but I had to stop it. And I got yelled at, but I was like, "These guys aren't thinking about personal income tax, okay?" Because they were twenty-two or something. They weren't thinking about saving money on taxes.

**STEPH**  So twenty million goes into the company, and it was a race to build this thing up as fast as possible. Again, there were four search engines that had gone public and Netscape, but there was still no Internet mania. People just sort of thought, Okay, you need to be a public company to get maximum credibility, to attract advertisers; it would be a great thing to do. So we knew that it was our mission to get public. And it wasn't an issue of raising more money, it was simply that it was expected and understood.

**TODD**   In the beginning of '97 most of those guys like Yahoo and Excite were trading under their IPO prices, and there was a huge question whether it was really going to work out. We still had a feeling, as Steph was saying, that we needed the credibility—especially in our ability to make acquisitions, which was something we knew we could not do with cash. But we knew that if we were publicly held it would be much easier to do with stock.

**STEPH**   So there was a race to get public simply for competitive reasons. If you get there first and you're the first in your category to be public, it gives you a head start: you get bigger, you get bigger faster, et cetera. We did everything we could from late '97, when the twenty million came in from Egan, and by spring of '98 we were ready to file. You can imagine how mad a rush this thing was. But again, at the same time, no one thought it was a slam dunk going public. Your stock would go down; there was no Internet sector, and there wasn't yet any Internet craze to go public. We were a little bit wary about it. What happens when you're public? You know, Todd and I have never run a public company, how would we . . . we were only twenty-four. It was all a little bit scary.

But by the beginning of '98 we had found a few banks that wanted to do this. To the big banks we were some small company, Internet-based; to them it was just Chinese. So we managed to find a couple banks that were smaller and more aggressive, and we went with them. Worked our asses off for the following three, four, five months preparing the S-1 and spending twenty-four hours a day down at our law firm, working like crazy to get this whole thing done. The market, by the way, was just continuing to climb. The NASDAQ kept climbing and climbing and climbing—and not because of the Internet, just because it was climbing.

**TODD**   We filed our S-1 on July 24, 1998. This is when the Internet hype really did start to hit its stride. This is when GeoCities went public—in the first week of August, I think. This was an exciting time. The day we filed our S-1 came at one of the highest points of the NASDAQ. What we didn't know was that in August it would start a very, very steep decline. And a very quick decline. And so during the August–September period, when we were preparing to go on our road show, companies were starting to pull or delay their IPOs.

So we said, "These clowns can cancel, but we've made it this far, dammit, we're going to go all the way and do this road show and raise this money." There was really a feeling of confidence, in view of all the statistics out there and all that we read in the press. So we went on this road show, and in fact the day we started we picked up the *Wall Street Journal* and on the front it said that something like one hundred and seventy-five companies had pulled and there were two on the road, and we were one of them.

**STEPH**    The NASDAQ was in the process of dropping down to 1700 points. One of its biggest drops since 1987.

**TODD**    It's a lot like right now. It is the same exact feeling—if not worse, maybe, because you had the huge Long Term Capital crisis going on sucking every country dry.

**STEPH**    We'd been well trained, and taught that road shows were the most painful thing you ever do. So we already had that built-in fear, and we're already on the road—ten meetings a day, ten days in a row, thirty cities. We were just going crazy. But the worst part was, as this market was collapsing, every meeting we went to people were looking at us, wondering, "Are you nuts? What are you doing here? You're never going to get our money. I don't care how good your business is, who your management team is, how great your product is; there's no way I'm losing millions."

Every meeting was like that, but Todd and I just persevered. We kept checking in with our bankers. "How's the book [tally of investments] coming along?"

"You guys better keep trying."

"How's the book coming along?"

"You better keep trying."

**TODD**    By the end of this thing we had maybe filled two-thirds of the book. Which is not good enough. Usually you're trying to fill double what the book is. Or more than the total sum of it, to create demand for the stock as well as protect you if someone decides to drop out at the last moment. It's not like it's a legal obligation if someone says they're going to put in the money; it only counts at the moment they put in the money. It was a very stressful point. And many of the people we met with were nice to us; they'd say, "We like the company. A month ago we probably would have invested in you."

They were very honest about it; it wasn't a personal slight to us. Nevertheless, we were spending a lot of time and money to put ourselves on this road show—not to mention the trauma of it emotionally—and to come out of it and to see that we weren't going to go public would have been awful.

We made a decision at the very end not to cancel the IPO technically with the SEC, but just to delay it and keep it out there. But it was extremely sad. You know, we had probably less than five million in the bank at the time, so we said, "Enough of this IPO stuff—let's get back to the business. We'll go into a recession, we'll downsize, we'll deal with it." We'd been in business at this point for already four years, so it's not as if we hadn't dealt with hard times before. But it was very sad—and the very worst part about it was that we had a company holiday party the following week.

**STEPH**   Everyone thought it was planned because of the IPO. So when we had the party, everyone was coming up to us with congratulations. And we were just like, *ughhhhh.* We didn't even have the heart to tell them we weren't going public.

We couldn't get the deal done at any price. We tried the price at eleven to thirteen dollars, and typically when there's great demand it goes to thirteen or higher. But in our case it couldn't even get done at eleven, ten, nine, eight dollars. Very quickly you realize, it doesn't matter how much you lower the price—and we lowered it—when no one wants to buy in. So the whole deal was dropped.

**JASON CHERVOKAS**   In the spring of '98, the Asian and Russian currency crisis tanked the stock market, and it looked like the entire IPO market was coming to an end; certainly the Internet mania was coming to an end. There were two local companies that were in registration and all set to go out: EarthWeb and TheGlobe. And I figured they were both dead in the water. TheGlobe got pulled back, and Earth-Web proceeded.

Bear Stearns was the underwriter on TheGlobe. They pulled back because they couldn't find a buyer at ten dollars, twelve dollars a share, or whatever the price was. In that range. EarthWeb forced itself out and got out to a whopping reception. Within a week and a half, Bear Stearns put TheGlobe back on the table for nine dollars a share, and obviously we saw what happened.

**STEPH** The week of November 7 another company went public—EarthWeb—and they went up higher. And then all of a sudden, within a twenty-four-hour period, the entire market changed.

Our bankers called us up on a Wednesday night and said, "Guys, you're never going to believe this. The phones have been ringing off the hook. Everyone wants into the deal."

We didn't even want to believe it, but as of midnight there was something like a forty-five million share demand for a three-million-share deal. Vastly oversubscribed. Now here was the bad part: The last time we had lowered that price, when we couldn't get the deal, it was down to eight bucks. So what happened? Our bankers left it at eight bucks. So instead of raising all the money we had hoped to raise, we were only raising a fraction.

We also had a couple other pressures going on. Our S-1 document, which is the only document that can allow you to go public, was about to go stale the following Monday. That meant we had to make a decision right then and there. Do we close this deal on Thursday night and price Friday? Or do we risk it and wait? The reality was, we couldn't risk anything. If the S-1 goes stale you have to refile and wait a month or two. If we were to wait any longer, what if the market collapsed and the window shut again? What if it was just a one-week window? So we said, You know what? Better to go public at eight or nine bucks a share than not at all.

So Thursday we finally decided to do it. We managed to get the price to nine bucks instead of eight, which was, you know, a nice little concession. But we basically locked down the price. It was a mad frenzy; we had to call everyone back up whom we'd basically told the deal was dead, and tell them it was up again.

**TODD**   It was surreal.

**STEPH**   There was no such thing as "hot Internet stocks" or "hot IPOs." Todd and I had real trouble giving away any stock for friends and family. A lot of people didn't even understand what an IPO was. Little did people realize what would happen the next day. We didn't realize what would happen the next day.

**TODD**   The next day, Friday the thirteenth, when we got up, we were very concerned. We hadn't slept; there were concerns that we were going to go back to war with Iraq, and so much bad luck had happened to us that we were pretty certain that something would chase this event.

But we did make it. We went to Bear Stearns in the morning, and we're sitting there eating muffins, and they're trying to keep us in a closed-off room. Eventually Ace Greenberg brings us down and he's doing his trademark card tricks and this type of thing to entertain us before it goes public. Eventually they sort of file us over to the part of the main trading floor.

**STEPH**   The deal was priced at nine dollars the night before; that morning, when we got there, they told us the deal was going to open at twenty, thirty dollars because of demand. Which was just insane. Unheard of—it had never happened in Internet history, in *stock market* history. Unfortunately, that kind of demand is only for the day of the trading; you don't benefit from it, you don't get the extra cash. So it was going to open at twenty to thirty dollars, which was . . . *neat*. It doesn't do anything for us, but it was an exciting event because it had never happened before.

**TODD**   We raised twenty-seven million dollars at nine dollars a share. The people who got the rest were the bankers' clients.

**STEPH**   So we found out it was opening that high first thing in the morning, which was insane. And by the time we were in Greenberg's office they come back in and tell us, "Oh, no, no, that was a mistake."

"Oh, of course it was a mistake."

"It's going to open at fifty to sixty."

*"What?!?"*

**TODD**   At this point we were both excited, but we were also pissed off. Because . . .

**STEPH**   People were obviously willing to put up hundreds of millions, and we weren't getting any of it.

**TODD**   This will be a story we will come back to many, many times, because it sort of followed us. Still, today, it's one of the main problems. TheGlobe fell from this high price.

**CLAY SHIRKY**   I reserve a special distaste for TheGlobe because that was the first model I could point to where it was created on a bedrock of nothing more secure than money. The myth at the time was that the Internet was simply a money multiplier. And that if you could invest two million and get four million, you must be able to invest twenty million and get forty. TheGlobe was just money-chasing in the most shockingly egregious way.

**SETH GOLDSTEIN**   It was just kind of bizarre, because you look at these companies like TheGlobe or EarthWeb—and they weren't companies; they were a bunch of people pretending to be a real company—and they either had really good bankers or a good investor, and they got packaged up and sent to the prom. You just kind of looked at that and were offended by it, but at the same time, felt like, "I can play that game." And that's what led to the Internet mania. The blind following the blind.

**BRIAN HOREY**   TheGlobe was a relatively small offering, so there was a supply and demand issue, and people at that point were unsure of what the growth prospects for the economy were. They looked around and thought, the Internet is the only place guaranteed to grow rapidly no matter what the economic situation is. And here was a deal they could participate in.

It surprised a lot of people, because the IPO market had been shut down for a month or two, and these were very young, speculative companies. That was really the wakeup call for a lot of people that the Internet would be an investment phenomenon. And I think a lot of it, too, was individuals who had started to use the Internet for shopping and news, and had decided they wanted to put their individual investments in these stocks as well.

**TODD**     It was a fantastic day. They're counting down until they open your stock for public trading—there's nothing like that moment. That was probably one of the most exciting single moments in our history. They're literally counting—*Ten! Nine! Eight!* The whole room went silent.

**STEPH**     I don't think Todd and I ever imagined, when we were just two twenty-four-year-old kids, that this would ever be happening to us. We were surrounded by all these guys in their forties and fifties, and I'd think that they'd seen this a thousand times before, but Todd and I were just like, *Whoa.*

**TODD**     I didn't have that feeling. I had the feeling that no one had seen anything like this in our group. That is, Egan, our whole management team—some of them were crying. This was an extremely emotional time. At any age. This was a damn exciting time. And it was the biggest. All the people on the trading floor talked about how they had never seen a run-up like this. They said that it was absolutely incredible to them. When we ultimately left the trading floor, we went down to NASDAQ, where they sort of toast you into this "new world," if you will.

Then we went back to the office, and we had some people who had been there for four years who had done extremely well, and we had people who had joined last week who just sort of have question marks popping in their head—What *is* this going on? But it was certainly exciting. Everyone in the company recognized that it gave us a new lease on life. Who knows what would've happened without it? We probably would've been forced into selling the company if we hadn't been able to raise enough funds in the six months after that.

**MICHAEL KAWOCHKA** Todd was basically regarded as the smarter one. He was definitely more hands-on. You'd see Todd there early, late at night. He seemed to give a shit more. We'd go get breakfast at nine-thirty and see Steph walking in at that point, didn't really see him there late nights, and see him walking around the halls, but it seemed like Todd was more of the brains of the operation. People regarded Steph, whether it was true or not, as the money connection. They both came from money, but Steph came from more money. His great-grandfather founded Nestlé.

**JASON MCCABE CALACANIS** The thing that got me upset was they were telling people they were the next Yahoo. That was their shtick, you know: "We're like Jerry Yang and David Filo." I knew Jerry Yang and David Filo, and they're no Jerry Yang and David Filo, not even close to as smart. They were just doing what Bo Peabody did three years earlier, making home pages—big deal. They weren't even in the top fifty of the Media Metrix, they were way down in the bottom. Their PR person was just overaggressive about how important they were, and I just creamed them. It was a bold move to make a prediction [that TheGlobe stock would plummet] when they were the highest IPO ever up to that point. They may still be the highest IPO on record for first day gain. Nine to ninety-one dollars a share is where they went in the first day—pretty staggering.

I knew they were posers. I went and met them; I met their company and got to know them, and got to know what they were doing. I said, I have to tell the truth—that's what I do, right? Let the chips fall where they may, and that's why people respect me.

**SUSAN BERKOWITZ** I had left to work at Wit Capital a month before the IPO. There was no lockup at TheGlobe, and I was fully vested when I left. I had a number in my head, X number of shares times X dollars—somewhere around thirty, based on the EarthWeb numbers. I'm at Wit Capital, and it's Friday the thirteenth. We have an eight A.M. meeting, and throughout the meeting an assistant keeps coming in saying, "Sue, the guys from Bear Stearns are on the phone with me. Indications are the stock's going to open at fifty. What do you want to do?"

I'm like, "I don't know. Let's talk about it later."

Twenty minutes later the phone rings again. "Indications are the stock's going to open at sixty. Seventy. Eighty. Ninety."

I became completely catatonic. I had no idea what was going on. That number in my head was now three times what it was. We were talking about a lot of money in the first place, but now we were talking about a *lot.*

Everything went off. I'm generally very composed. Ron Readmond [Berkowitz's boss at Wit Capital] comes into my office and goes, "Are you out?" He was an army guy, very stern.

I said, "I don't know."

He said, "Get those motherfuckers on the phone now!"

So I call the guys at Bear Stearns, and start by saying, "Hi, it's Sue Berkowitz. I have the former COO of Schwab on the phone, and he wants to ask you something about my trade."

And he goes, "Is she fucking out yet?!?"

The stock had been open for about five minutes.

He says to them, "What the fuck are you doing standing on the goddamn phone with me? Get down to the trading floor and sell. And by the way, do not call her when you're done. You call *me.*"

He's screaming, I'm crying, I have no idea what's going on.

That phone call probably saved me two million dollars.

There were only six people or so at TheGlobe who made a significant amount of money—the core people who were there from the very beginning. I remember thinking, "Oh, I'm an Internet millionaire."

And Andy Klein said to me, "Are you going home tonight?"

I was like, "No, I have more work to do."

He said, "You're an idiot." It was a Friday night, and I was still working. But I didn't have to work anymore.

I went over to Dewey's Flatiron, and Ellen Gold—a woman who worked for me at TheGlobe, who is really nutty—sees me, comes flying out of the door, leaps into my arms, and knocks me down onto the street. Screaming. They were all so drunk, and it was really a hilarious night. It was nuts.

**STEPH**    We learned a lot of things from that IPO. Number one, it was exciting that it went up a thousand percent; it gave the company a huge amount of visibility worldwide. People were calling us from

Switzerland, where I have family; people were calling us from India just saying, "Haven't seen you guys in ages, but what the hell are you doing on CNN?" You turn on the international news, and there's Todd and Steph. It's just like, *What the hell?* The flip side was, of course, that suddenly expectations were sky-high; and people were telling us that day, "You're never going to maintain that price. It has to come down." Which it did, naturally.

It also meant another thing. Which is that after that day there was a sudden, new, worldwide Internet euphoria that just happened out of the blue. And unfortunately for us, what followed was that everyone wanted into every Internet IPO. If you just said "I'm a dot-com," people were flooding you with money. And it meant that all of our competition were raising a hundred million dollars for the same percentage of the company we had for twenty-seven million dollars. And the reality is, you have to compete with those companies. So very quickly you could see what was happening. We had twenty-seven million dollars and the stock was coming down; they've just raised one hundred million dollars, and their stocks were going up. We had to compete with them, and six months later we realized the only way to really do that effectively was to go do a secondary offering.

It was an interesting moment; IPOs are the roughest period, and you only want to do it once in your life, but there we were six months later going, Okay, we have to do it all over again. Back on the road.

Todd and I were just like, Oh my God, we don't *want* to do it again. But we had no choice.

**CLAY SHIRKY**  Jason Calacanis and I had a public bet that The-Globe would fall beneath thirty dollars within some number of days of its IPO. And we both disliked TheGlobe, because we both distrusted the business model. It seemed purely cynical manipulation. I never saw them do anything where the message wasn't "We're just like GeoCities, only better, because we're us." There was nothing there. And indeed it went below thirty dollars a little bit one day, so I owed Jason lunch at Nobu.

**JASON MCCABE CALACANIS**  I made a public bet that The-Globe.com would be trading below thirty dollars within thirty days of their IPO, when it was at ninety-one. I bet anybody sushi lunch at Nobu and dined out for one hundred and seventy five dollars [on Shirky] at Nobu for lunch, and it was very funny. But I made the bet on the WWWAC list in front of everybody, and I wrote an editorial about The-Globe, saying TheGlobe is a farce. It's a scam. The word I used was *scam*. And I really cannot explain why. You know, their stock is probably trading today at fifty or seventy-five cents. [Checks the market update on his computer.] Yeah, less than fifty cents.

It's the biggest crash in all of the Alley, clearly—probably in the whole Internet industry. It hasn't hit zero, so when it hits zero, it will be. But I knew it from day one because I met the guys and I realized that they were idiots and they were just backed by people trying to make quick money. I had to make a stand. I have to say the good and the bad, and this was bad, and I made my stand. I think it became very popular to beat up on them after I did, and I think I ruined their lives to a certain extent.

**JASON CHERVOKAS**   TheGlobe was a business that never was never really a business. It was a college hobby. It got money from a private investor who probably didn't really know what he was investing in; he managed to seize this moment to get his money out, and clearly what killed TheGlobe from an investor perspective was not so much this crazy first-day IPO but the secondary offering that followed. It was shortly thereafter that the stock split—a secondary that included a lot of insider selling—and it just absolutely destroyed investor confidence in the business, because it seemed that it was just a way for people to cash out. That really destroyed the credibility of TheGlobe.

*On May 4, 1999, in the secondary offering Stephan Paternot sold eighty thousand shares of TheGlobe stock at twenty dollars per share, reaping a total of $1.62 million. Todd Krizelman sold one hundred and twenty thousand shares at the same price for a total of $2.4 million. Investor Michael Egan sold two million five hundred and twenty-three thousand nine hundred and forty-two shares for more than $50 million, invester David Horowitz sold twenty-eight thousand shares at twenty dollars each for $560,000.*

**JASON CHERVOKAS**   TheGlobe was a business that was run by two very nice young guys who probably were okay for starting the business, but had no business running a public company. No one was really at the reins to say, Okay, we're going to change management now. We're going to actually make this function as a business. By the time they did that, the horses were out of the barn.

**TODD**   You know, a lot of things had changed since we'd gone public. The Internet "industry," if you will, had changed dramatically. Not only were people putting a lot more money into new companies going public, but in many ways that was the least of our problems. Suddenly every employee who had an idea thought they could go raise ten million dollars, and in many ways they could. Average ideas would get very well financed, and while you knew that they would ultimately go out of business, it would take some time. You're seeing that now—a lot of those companies diving. But I think thousands of companies got financed, so that's going to go on for a while.

**STEPH**   I think the Internet's own euphoria and hype became its own worst enemy. Our own spark became, in a sense, our own worst enemy. I think now everyone's paying the consequence in the Internet space for what happens when too much floods in too fast.

   And of course the biggest problem with this Internet euphoria is that everyone who entered the industry after November 13, 1998, went in with dollar signs in their eyes. That was the spark. "I want me a net comp'ny cause I kin make a billion." That's what it was all about. And that was the biggest danger. People going in for all the wrong reasons. Employees coming in for the wrong reasons. And they'll tell you "I'm here to build the company," but you've got to be careful. You end up in a situation where there's a lot of people chasing after the gold nuggets out West. This is a culture that ultimately can't survive that way. And what's going on now is a cleaning out of the whole system.

**TODD**   We called it an entitlement syndrome. We had many employees who would come in monthly asking for more money. We'd usually just tell them to go to another company, and you'd have people who would say—I love this: "Why haven't I made a million yet?" I said, "Do you realize the president of your company hasn't had that opportunity? Your CFO hasn't had that opportunity. These guys are seasoned executives." We reward people on merit, we do all of the competitive surveys to make sure we pay well, but it is unbelievable the way the press fueled their expectations.

Then there's the personal level, which was extremely frustrating. Steph and I would read how we were worth *sooo* much money on paper. In the very beginning Steph and I agreed that we wouldn't change our lifestyles. We still live in the same apartments we bought before the company went public; we lead basically the exact same lives we've led for the last five years.

**STEPH** And thank God we didn't change our lifestyles. You know, we went from zero to one hundred million—which we never had—back to zero.

**CLAY SHIRKY** Their one-day slide, from the launch price down, was I think the largest transfer of wealth from individuals to institutions in the history of the stock market. It was just surreal. It was so oversubscribed and so bid up that it was essentially a voluntary tax paid directly to financial institutions by investors who'd gotten the Internet fever. And TheGlobe was, the last time I looked, trading at something like five-eighths. And I can't think of a nicer bunch of people it could have happened to. Everything they have ever done has been cynical, and I have never detected any interest in actually using the Web to do anything other than make money.

**JASON CHERVOKAS** The problem with TheGlobe is that the perception is just that it's damaged goods, so it's really hard for them to find an exit. I know any number of businesses in New York that have looked at acquiring TheGlobe, and all of them have passed. There have been a lot of different reasons, but one thing that carries over in every case is: If I do this, my investors think I'm an idiot, because there's just such a stink around TheGlobe.

**STEPH** I think we have a great company; I think there are a lot of great companies out there, despite where their stocks have gone, and they'll make a comeback. They'll make a comeback this time, the way all companies should have done it in the beginning, which is through slow, steady growth.

There are no guaranteed bets. You never know where it would've gone, and certainly every decision we made we based on the best knowledge we had, the best advice we could get, and what we thought would provide the best return for investors and fulfill our dreams. So you can never say you made all the wrong decisions.

**TODD**   We had offers in year two to sell the company, and God knows we would've made a lot of money.

**STEPH**   If we had sold out in the second year we would've never experienced that mania, gone public, done any of those things.

**TODD**   Knowing now who I am and who Steph is, it would be like losing a piece of me. You know, money's great, but it's only great if you have it with something that you love doing.

**MICHAEL KAWOCHKA**   When the stock was at four dollars, we were like, It's not going to go much lower. Then it went to three, two-fifty, one. We had a bet in the office—is it going to go below a dollar? We didn't know. Then it was like, Holy shit, it's ninety-six cents. At one point I said to a friend, "Maybe I should buy a hundred shares. It's like a night out—why not?"

He said, "A hundred shares? I'd rather buy a hundred tacos."

Then the running joke every day was, a hundred shares or a hundred tacos—who's better off? You'd be better off with the tacos. We'd go out to breakfast every day and be like, Wow, my breakfast is four bucks. I could buy five shares of stock for that.

**CLAY SHIRKY**   Anybody who lost money investing in TheGlobe can certainly look at the original investors and say, These people were like charlatans. There was nothing here. There was nothing here when they started it; every press release issued about TheGlobe in the early days was all about how the [Alamo] guy gave them twenty million dollars. Anyone who was taken in by that—particularly after that myth was solidified into the notion that this somehow constituted a legitimate business model—could rightly feel upset and somewhat betrayed. There weren't any entrepreneurs at TheGlobe. It wasn't about their business model. It was spinning money out of thin air.

**FRED WILSON** Where they are today—it seems they're kind of lost.

**STEPH** We can look back now and know next time which of the steps we might want to take differently. We can know what feels bad, what feels good; you become infinitely more mature.

*In January of 2000, under the pressure of a depleted stock price, Paternot and Krizelman announced they would be stepping down as cochairmen of their company.*

# FUCKED COMPANIES

*On March 10, 2000, the NASDAQ hit its all-time peak of 5,048. A lot of people had gotten very rich during the years of plenty on Wall Street.*

*But in late March, a bearish* Barron's *reported that CDNow, an online retailer, along with more than two hundred other Internet companies, were running out of money. With a merger between AOL and Time Warner pending, analysts began to realize that the valuations of Internet companies were way out of step with the rest of the market. On April 10, and for the four trading days that followed it, the NASDAQ lost between one and three hundred points each day. On March 12, 2001, the NASDAQ would drop below 2,000 points, just a year and two days after its height.*

*The shakeout had disastrous effects in Silicon Alley, as the most prominent companies were punished by the Street. First went the e-commerce sites. Then the community sites. Then the content sites. Then, with the collapse of the war chests of most of the companies surrounding them, the high-flying interactive agencies. Razorfish and Agency were clobbered, losing over ninety percent of their value by the end of the year.*

*Massive downsizing followed, as companies abruptly closed or cut huge chunks of their staff. Published estimates put the losses in the Alley at about fifteen thousand people, but to anyone working in the industry it was clear the real numbers were far worse. Virtually every company doing business in the Internet was involved in a desperate battle for life.*

*Disaster became so common that a website called FuckedCompany.com appeared serving as a doomsayer for dot-coms everywhere. The Dotcom Scoop list, a wireless messaging community on Upoc.com, gave up-to-the-second reports of layoffs of more than three hundred workers at DoubleClick, two hundred at Agency, five hundred and twenty at Razorfish.*

*Perhaps even harder hit than the public companies were the thousands of start-ups that had yet to go public, which were often suffocated when angel investors, many of whom had made their fortunes on the tech bull market, could no longer provide new investment money.*

*Even experienced entrepreneurs like Scott Heiferman—who had success-fully sold his first business, i-Traffic, to Agency.com—found themselves inca-pable of raising additional funds to keep their companies afloat. He financed, developed, and distributed 250,000 free copies of the RocketBoard, a keyboard with Internet buttons, but closed the company before filling more than a few hundred paid orders. Spotted back at i-Traffic a week later, looking glum and out of place, he was asked by an employee what he was going to do with all the unused RocketBoards. "Use them for doorstops, I guess."*

*A wireless device called Modo.net, which provided an updated city guide, made its Flatiron Partners money last less than two months. For months after the company went out of business, its advertisements still decorated billboards and bus shelters around the city.*

*In the months leading to April 2000, everyone in the industry knew a cor-rection was inevitable, though. But for years the evangelists of the new economy had found themselves proven right again and again, and their belief in the infal-libility of the tech market remained steadfast. Venture capitalists continued to fund new start-ups because investment banks could still sell them to investors in the public market. It didn't seem to matter whether profitability could ever be achieved. The analysts at the investment houses continued to encourage their clients to buy, unwilling to believe that the Gold Rush was truly over.*

*But the NASDAQ crash of 2000 left thousands of people without jobs and many more with the hopeless sense that the dot-com dream had really ended.*

**ALICE O'ROURKE** I guess we should have seen the end coming. One of the signposts should have been when I heard a VC say, "We're not going to invest in anyone over twenty-six years old."

**SETH GOLDSTEIN** Silicon Alley was built on hype. It was a bunch of people who were really good at telling stories and got a lot of people to listen, and in the process built companies that capitalized on that buzz and that enthusiasm and momentum. But there was no core asset other than the momentum.

**FERNANDO ESPUELAS** It's painful to see ninety percent of your market cap[italization, or value] disappear. Before, we were very open with our team, but when you're public you can't risk the leak. People see stock price as a daily report card, and it's not.

**JASON CHERVOKAS** Everyone thought they were going to be Bill Gates, Barry Diller, and Rupert Murdoch rolled into one, without even beginning to understand what that actually meant. People were very naïve about what they expected, about what the financial opportunities really were, and what revenue could really exist. I think a lot of stuff caught people by surprise, a lot of things were unexpected. And I think, more than anything else, that there was a sense that something revolutionary was happening, and therefore no traditional rules applied— which meant people were free from actually doing traditional analytical thinking. As a result, we did a lot of crazy things.

**JERRY COLONNA** We had an entrepreneur who came in here and told us the reason he started his business was that two of his classmates from business school had started businesses, and they were rich and he wasn't, and he's smarter than them, so it was his turn. And that's a lousy reason to go through all the pain of starting a business.

There was a sense of entitlement. "You got yours, now it's my time to get mine." And I think it was devastating. I think we are reaping the rewards of that attitude now. I've talked about the resentment we see in the press turning to glee. A lot of the backlash you see in the press is coming from Alley people being snotty. There was an attitude of "We're king shit, and we're going to take over the world," a lot of which was fed by the paper millions.

**ESTHER DYSON** There was a sense that if you're young, you must be smart. And there was really incredible economic ignorance. You don't need to be an economics professor; you *do* need to know that stock markets go up and down, and how the fundamental rules of economics work.

I've been amazed, not just by Silicon Alley, but by the level of people's ignorance or willful blindness. I wouldn't blame any particular group. It was mass-delusionary self-gratification. Everybody feeding on everybody.

**SCOTT HEIFERMAN** When you're building a company, you're running toward a cliff. You can't crawl or walk, you just have to be running. It's like a plane taking off. You've got to run to take off, and you're running toward a cliff, and there's either going to be a bridge there or not.

Money is the lifeblood, and if we're out of money, we're out of business. So that's the plain and simple of what happened to RocketBoard. The business was going well. I worked as hard as I could. I've basically taken a bit of a leave from i-Traffic for the past six months to make it happen, and I gave it my all, and I couldn't pull it off. We ran out of money, couldn't raise more. Raised five million bucks, but couldn't raise more.

**KYLE SHANNON** We fucked up. We got seduced into thinking that a slick pitch for a screenplay is a good movie. There were a lot of people pitching slick screenplay ideas, and a lot of people funding movies, for people who never made a movie before, didn't have a screenplay to back up the pitch, didn't know what they were doing, and didn't know anyone who knew what they were doing.

Ultimately investors will believe in the potential of what's out there, but we shot ourselves in the foot with all the companies that weren't really businesses, just a bunch of yutzes that got together with ten million dollars to blow. They blew through their ten million dollars, and I guess it's probably because we live in a world of such perfect information that when one goes, they topple like dominoes. Then the word gets out immediately, and suddenly "e-tailing" goes from being the six-month word of the day to, "Oh my God, if you're an e-tailer you're in trouble."

I think it's completely our fault. I think that of the companies that died, there were probably some on the tail end of this that were good companies, but needed some funding that they're not getting because the attitude is, Fuck it, you're all dot-coms, you all suck. So there's probably some unfortunate things in there. But I would guess if you went through the graveyard, ninety-three percent of them had no right getting funding.

**JASON McCABE CALACANIS**  It's been a tough few months. If you look at the NASDAQ chart and the growth of the NASDAQ—which is probably the closest thing you can have to a sort of metric for technology and stocks and growth—it does something along the lines of a very steady growth curve for ten years; a huge spike; a huge dip. If you take out the huge spike and the huge dip, that's probably where it is supposed to be.

The NASDAQ should never have gone to 5,000 as quick as it did. It probably shouldn't have gone to 4,000 as quick as it did. If you want to get an idea of how quick that actually happens, I'll give you the sort of loose snapshot. It took the NASDAQ from '85, '86, all the way until 1996—ten years—to reach 1,000. It took the NASDAQ from January of '96 to some time in '98, '99, to go to 2,000. But then it went from 2,000 to 3,000 in '99, in about six months in '99, and then it went from 3,000 to 4,000 in a quarter. And it went from 4,000 to 5,000 in a quarter. And then they came crashing back down from 5,000 to 4,000 in a month, and then it came crashing down from 4,000 back down to 3,000 over a couple of months. So basically, with the NASDAQ cruising in the 2,700, 2,800, 3,000 range, we are back where we hit in the fall of '99, which is when this thing got out of control anyway. So you basically have this last eighteen months of insanity. Everybody knew it was insane.

**MARISA BOWE**  I got all these calls from reporters when the NAS-DAQ crashed, looking for some shock quote. And I said, "It doesn't put any more money in my pocket if someone else loses some paper money."

**CRAIG KANARICK**   The king of Sweden was in town, and our design director in New York is Swedish, and got a private audience with the king. He got, like, an hour of training in how to bow. He goes in, and the king says, "So you guys just laid off a lot of people in your Stockholm office at Razorfish. How is that going to affect the economy?"

**STEFANIE SYMAN**   There was some pride in being scrappy and making it in the early days, but by '99 that was just nowhere to be seen. It was, "Let's staff up, let's hire three hundred people, and let's do this website." I think if you were established in the early days you had a very recent memory of having to make something out of nothing, and you had perspective for that reason.

Unfortunately, some good ideas have basically failed because there was too much money. The ratio of money to ideas was all out of whack because of the possibilities. On the one hand, it's true that a lot of ideas were incarnated or tried out, but in many cases it would have been better if the stars weren't in everyone's eyes and people had approached it with some of that caution that was appropriate, and very present, in the early days.

**NICK NYHAN**   I think the Internet and Silicon Alley turned eighteen this year—where you're no longer a kid, you're an adult, you can vote, we can send you into a war, and you have to come back and be accountable for your actions. You're no longer a teenager who can blow money. And I'm thankful for it. I think it's going to make it better.

**CRAIG KANARICK**   I never really believed in the money, and I still don't believe it—which is good, because it's almost all gone now, thanks to the stock market. Jeff and I haven't sold a single share. Not one. So all that stuff was always paper. We got a little money from early investors, but not a lot.

**JERRY COLONNA**   It's really hard to start a business in the United States. In 1998 I wrote a column in the *New York Times* where I said the Internet bubble is too big, and we should let the air out. Because, among other things, there were going to be devastating psychological effects for the entrepreneurs. And one of the things I said was, I knew an entrepreneur who was close to selling his business for twenty-five million dollars, and he felt like a failure. After four years of work, he built the business to a couple of million in revenue, he was selling it for twenty-five million dollars, and he felt like a failure.

We've totally altered our perspective. My father toiled away for thirty years with a company, and the company closed on him. [My partner] Bob Greene's father was a failed entrepreneur four times over. And for someone to spend four years and walk away with a business worth twenty-five million dollars—that's a success in anybody's metric. But we had this weird time frame, from '96, '97, and especially '98 to '99, where somehow the measure of success became the number of billions of dollars in paper value you created. And that's insane. That's unsustainable.

**DAVID LIU** The market is just punishing all of us. This is completely scary, because it could completely wipe out someone like [TheKnot.com], too.

**SCOTT HEIFERMAN** People who got involved with the Internet early were going against the grain, going against the prevailing themes, and that personality keeps that mind-set today. Although, you could ask, is there a burnout factor? I've had conversations in the past few days—maybe it's because I've had this experience of having a company fail in the past couple of weeks and the ensuing conversations around that, people paying their respects or whatever—but there's some really down people right now, who are normally the big cheerleaders.

**SETH GOLDSTEIN** I think it's pretty grim. A disproportionate number of people have finished what they're doing and are kind of sitting on the sidelines, figuring out what to do next. Some of them have more time on their hands than others. A lot of people are reading. A lot of people are trying to reconnect to what makes them feel good, because there's no quick buck, and how hard you work doesn't really determine how much money you're going to make right now, so people are questioning all that. Equity doesn't matter—cash matters.

**CRAIG KANARICK**  The stock market has been completely irrational for a couple years in both directions—just like the press. Having said that, I don't envy anybody who is trying to figure out where to invest in the current marketplace.

What people in the digital industry came out and said was that the old ways of evaluating companies would be irrelevant; there are no rules. But now we want to say we're unfairly valued—so that means there *are* rules. I don't envy anybody trying to figure that out.

I don't expect anybody to figure the difference between all the companies competing in the same space. In any new industry it's difficult to tell who has strengths and who has weaknesses, especially with industries based on future growth. Do I think our company is worth more than one times our revenue? Do I think it's worth more than a hundred times revenue? Probably not. All of that is a bet on the future.

People are taking a very shortsighted view of what is going on. If they have even a bad quarter they demand that the company be cut in half. That's like saying that if a baseball team loses six games in a row they should be kicked out of the league. For whatever reason, people want these businesses to function at an impeccable level. Twelve weeks [a financial quarter] in the business world is a really short amount of time. We planned on being here for twenty or thirty years. If we decide to make a change to get our shit together and it takes twelve weeks, that doesn't mean we're worth half as much. I think it's absolutely a crazy time.

**ESTHER DYSON**  There's this wonderful story—it's an economists' joke—about a little boy and his grandfather. The boy sees a ten-dollar bill lying on the sidewalk, and he starts to run to pick it up, and his grandfather says, "Leave it alone. The market works. If it were real, somebody would have already picked it up."

The current version of this is: A banker is trying to evaluate a company, and this other banker comes along and says, "The VCs paid eighty million dollars for it; it must be worth at least two hundred million dollars." And neither of these has any justification. You've got to look at the asset itself. Investors were investing on the basis of their expectations of other investors' behavior, not their expectations of the company's behavior.

**KEVIN RYAN** It's very hard to value early-stage companies. What really happened was that the life cycle of companies was pushed forward, and that's not necessarily a good or bad thing. What used to be third-stage venture capital was replaced by an IPO. And so that's much higher-risk and much higher-reward. Eight years ago, if you invested in a company and they went public, the odds of that going up ten times in two years . . . it just didn't happen.

So the good news is, if you bought Yahoo or DoubleClick or AOL or something like that, you made a tremendous amount of money. Now, you can't do that without the trade-off, which is that four of the companies you bought that IPOd end up going out of business, or selling for a very small amount of money. Either you understand the risk and you play, or you should stay out of that market.

I don't think it's necessarily a bad thing, but if people are going to draw a conclusion from that . . . the same math has always applied to venture capitalists. No one gets worried that Kleiner, Perkins has tons of companies that go under. They have enough winners that it works out fine. I actually think most institutional investors more or less understood that, and have done pretty well over the last two years. Just with a greater variety of returns.

**ALICE RODD O'ROURKE** Everyone was buying in, with a wink: "We will get these businesses going, and they will continue to generate a loss until they own the market." And that's what I think was in play here. There were companies that could have stopped growing, and could have created a profit, and would have done so based on good management. Whether highly experienced or relatively inexperienced, everybody bought into the proliferation idea: "We are going to keep spending as long as someone is willing to fuel that growth."

**DAVID LIU** I think the pendulum has swung far across the threshold of what equilibrium is. I've used the phrase "the class of '99": any company that got its initial financing in 1999—and I'll go out on a limb here—was categorically overvalued. They raised too much money on too high a valuation. I don't care what category you are—fiber-optic networking, B2C, B2B—if you raised money in the year 1999, you were overvalued.

Boo.com is who everyone cites as a train wreck. They shouldn't have had the opportunity to spend so much money, so their valuation must have been out of whack. I can tell you that our online competitors just merged and earned a valuation of four hundred and fifty million dollars; four hundred and fifty million dollars, when I'm trading at around fifty million, is insane, but the fact that that was done is an indication of just how out of whack things had gotten.

I think that iVillage trading at a valuation of over a billion dollars was ludicrous at the time. I think the fact that people believed they were worth that much only added to that. TheGlobe was also a travesty; these people were showing hardly any revenues at all, and they weren't showing a path toward a profitable business model. We started joking that there was a new valuation model, as opposed to P/E ratios or something like that—a head count. Because there was no other rhyme or reason for what was going on. TheGlobe was the worst one when it hit the public market.

**DOUGLAS RUSHKOFF**  The advertising agencies—all their interactive divisions are shutting down or shrinking down. There are these booms, when billions of dollars pour into NASDAQ and it has to be spent on something. You can only spend so much on routers. And it gets spent on people. A way to get more investment dollars in your company was to have more bodies. Your valuation could go up a million dollars per body—that was before burn rates were taken into account. You wanted more bodies.

**KYLE SHANNON**  I just saw this thing on CNN where forty-five companies have closed their doors in October and November [2000] alone. I assume public companies—well, maybe not, but all dot-com companies. All because they didn't get second and third rounds of financing, because no one's financing that shit any more.

I think in the long run it's actually a good thing. It's kicking the snot out of our stock price, but what it will do, hopefully, is put the fear of God in people's hearts, to say "Maybe you should get people that have an actual idea and an actual business model that actually fit together, and are bright enough or have the management skills, or whatever that magical combination is that can actually create a business." Rather than what was happening before, which was that literally anyone who could write a business plan could get funding. We had three employees in a three-month period get funded for their ideas. One got ten million dollars, one got twenty million dollars, and one got sixty million dollars. My response was, Your first sixty million and you're out of here? What kind of loyalty is that?

The guy who got sixty million was one of our brightest guys—he's doing well. The ten million and the twenty million? Those companies are both dead.

**GENE DEROSE** I think who really tended to get screwed were middle- and lower-level employees at many, many of these companies, who just didn't have a clue what they were getting into. It's just hard to know whether they got sold a bill of goods by anyone, or had a neat little experience, or whatever else.

The people for whom I long had a kind of amused disdain were the really, really qualified corporate types who were young, reasonably entrepreneurial, but could never quite pull themselves out of the safety of these places to get involved—and then in the last two years rushed into it when conventional wisdom declared that this *was* changing the world, and millionaires were being minted by the hundreds . . . and it was exactly the wrong time.

**SCOTT HEIFERMAN** There's never before been unemployed people in Silicon Alley. The idea of not being able to go across the street and get a job—in this little short history of five or six years, this is the first time that's happened.

**DOUGLAS RUSHKOFF** When I came out of college it was impossible to get a job. That's why they called it the slacker generation, or Generation X, and all that. It was literally "Go find a temp job." I tutored kids on their SATs for a living for three years because we were in a recession.

A lot of kids coming out of school now demand sixty, seventy, eighty thousand dollars a year, because they know HTML—which they think is coding, and it's not. They think it's a computer language and it's not. It's like word processing. It's nothing. Most of them were given what turned out to be worthless shares in companies, though they did get salaries.

These kids are in for a shock, because the so-called skills they had are quickly being replaced by programs like DreamWeaver, and the only people who actually know database structure, or C++, are growing up in Bangalore and not here, and get brought in on H1B visas, and paid twenty-five dollars an hour that they send back to their families. And as these companies close, there's going to be more and more kids. Now, finally, I'm hearing executives say, "Kid wants 60k to come in and be an assistant HTML guy—who the fuck does he think he is?"

They've got another think coming, they really do. They're smarter about certain things, but their expectations are so high that it's going to be hard. They're going to experience at twenty-eight what those fifty-five-year-old businessmen experienced in 1979, when downsizing happened and these guys who were making $150k a year had to go look for jobs and had to go to these rehabilitation centers to get new résumés. It's going to be hard.

**KYLE SHANNON**   I have a degree in acting, so I have more money now than I had ever planned on having in my entire life combined. So personally, I'm fine. Where it is, it is. If it goes up great, if it goes down . . . if it goes down too much more then I'll probably be in debt, but hey. Shit happens.

What I think is very ugly is that a lot of people left good careers, good jobs, for a handful of options that were worth $100 million for three days and are now literally not worth the paper they're printed on. In some cases the stock is so far underwater—the option vesting price is up here, and the stock price is down here—that the chances of them ever being worth anything are slim to none. It's very tragic: people thought they'd come in, do their thing, retire, and fish for the rest of their life. And that's not the case.

*What are some creative ways to stay afloat in these*
*difficult times?*

**JEREMY HAFT**   In order to extend your payables, instead of not paying your vendors, you send them a check with the decimal point off by one. So if you owe a vendor one hundred thousand dollars, you write a check for ten thousand dollars. At the end of the thirty-day period. So then it takes them fifteen to twenty days to process the check. Then they call you up: "Oh, we're off by a decimal point." Then you say, "Oh, God, sorry. Accounting error." They send back the check, and then you do what my grandfather used to do in the garment business—you send the check again, but you don't sign it. So you can extend your payables for up to six months. Some of them don't pay their vendors at all.

**SETH GOLDSTEIN** No company's gone out of business with money in the bank, so it's about stretching your cash. If the five million dollars you have is expected to last a year, make it last two years. Then you have to deal with the layoffs and hard decisions.

**KYLE SHANNON** A lot of the services companies in our sector were heavily dependent upon revenue from dot-com start-ups. When the market shit the bed in April, it screwed up the dot-com market. So companies were burning through, say, five million dollars a month, and they had maybe three months left, but they thought they had a second round coming. Then April happens. *Oh, fuck.* They realize that second round may not come in, but . . . *it will be fine.*

Well, it doesn't come in. So not only do you lose the work going forward with that dot-com, but you probably don't get paid, because they declare bankruptcy. They say, We know you did X million dollars of work for us, and God it was good, but we're going to have to replace it with a little black page that says "Sorry, we're gone." Hope you're okay with that.

**CRAIG KANARICK** If the company isn't doing well, you are going to be holding the bill in a couple of months. We've had a couple of clients who were working on a strategy piece: "Here is what the strategy is, and this is what it's going to cost." After a two- or three-month period, their stock gets cut in half. They still want to do the project, but their investors will kill us if we tell them we're investing eight million dollars and they want us to close a factory. So we have to wait a quarter before we get back on track.

I think it's fair for investors to say we should be able to predict our business cycle. We absolutely knew that the world was going to change. We used to complain how people used to come up to us and say, "We want a website. We don't know why, but we want one." We knew that once everybody did that, they would say, "Why *do* we want a website?" Then, once everybody had one, it was a matter of upgrading them or applying to the outside world.

We knew business was going to change; the problem was, no one expected it to change as fast as it did. Should investors be upset that we didn't predict our business cycle? Absolutely. Should they be this upset and this desperate in this situation? Not at all.

We sold them on fear, and that's how we got all that money in the beginning. Now we're selling them opportunity, and it takes longer to make those decisions.

**SETH GOLDSTEIN** Every company that was built in Silicon Alley depends on some other company that was built in Silicon Alley. Now there is a very painful process of self-examination: What were we thinking? Why did we make those decisions? How do I learn from those decisions? And what do I want to do next with my life?

**CLAY SHIRKY** Many of the community sites are effectively cultural institutions, which should be run potentially for revenue but not for profit. Particularly not for the sort of profit that generates the IPOs that generates the VC interest.

Many of these businesses are unable to find alternate ways of structuring access to capital, other than "We're going to get ten of you together, and we don't mind if nine of you fail, as long as one you pays out twenty to one."

The changes in the Flatiron [Partners] portfolio have largely come about because VCs can only lose one times their money—so the worst thing to do with a failing company is try to help it out, because that raises the amount of money you might lose. So the end, when it comes for a lot these companies, is much swifter than anyone imagined, because there's no way to capture the revenue stream without chopping off heads, and heads are where the brains are.

**OMAR WASOW** The good news is that it's been wonderful to see, in the last eighteen months, people taking black entrepreneurs and concepts targeting African Americans seriously. That's been great, that's a wonderful thing. That's progress for the Net; it's progress for the world. Just as Viacom acquiring BET validates the value of media companies serving this target audience as opposed to dozens of others, it is a really great confirmation of the importance of those media companies. So does the fact that venture capitalists are backing these companies and other private investors. I think it's a great certification or vote of confidence.

I'm enthusiastic about the fact that there's money sloshing around. As with the broader Internet, though, there's been a lot of dumb money backing poor concepts with weak management. And that would be fine if there weren't a broader syndrome that's endemic to prominent African Americans everywhere—which is that when it happens to one, it becomes a question about everybody. When Modo.net goes under six weeks after launching, people don't say, "This wireless thing is looking shaky." But when a company like UBO [Urban Box Office] goes under, people say, "This urban thing, this ethnic thing, doesn't seem that credible." It's unfortunate—and it's not UBO's fault specifically, it's the broader issue of how people look at this market. People confuse the weakness of particular companies with a weakness in the market. And the reality is, there are specific companies that are bad even when the overall opportunity is huge.

**KEVIN O'CONNOR** There are probably more opportunities for start-ups now, because you have a huge number of people using the Internet. When we started there were only ten million people using it. There are just different types of opportunities.

I think the ignorant euphoria that anything that moves will be a success is gone, as it should be. That never should have happened. It was absolutely obscene. My big fear was that the Internet was going to become like the savings and loan crisis—that a few bad apples would ruin the entire market for everybody.

When I got out of school there was massive unemployment, and no one got jobs—there just were no jobs out there. But recently, when people got out, there was no such thing as unemployment. If you were unemployed in that market, there was something wrong with you. Some people got out, and there were unbelievable amounts of capital available to them. But things can turn very bad very quickly. My fear is that we have had it too good—that it has all been too easy. This company hasn't suffered a trying moment or a tough time, and it is going to come because it always comes.

**JERRY COLONNA**   Lots of good companies are being unfairly punished, and not just the ones in our portfolio.

**FRED WILSON**   Look at DoubleClick. DoubleClick's stock has gotten creamed; people think that there's no value there. But that's a great business they've built.

**JERRY COLONNA**   It's a great business. It's a market leader, it's an innovator, it's a pioneer. It's done everything right. The one thing they did wrong was the way they handled the press announcement on the Abacus acquisition.

**FRED WILSON**   They didn't anticipate the privacy issues on the Abacus deal correctly.

**JERRY COLONNA**   I don't fault them for making the acquisition, because it was a great acquisition—perfectly logical. There are companies out there competing with DoubleClick that are far worse violators of our collective privacy on a continuous basis that aren't nearly as punished as DoubleClick is. And I think that over time people will see that DoubleClick is a phenomenally good business, managed by really bright, capable people.

**FRED WILSON**   There are a bunch of companies out there that have gotten slammed because of the association with the Internet—that are doing just fine, and meeting the numbers, and growing nicely. But right now, anything that has to do with the Internet and Internet economy is very suspect on Wall Street. And that's okay. What's going on is, the portfolio managers on Wall Street are saying, I don't want any Internet in my portfolio right now, and I'll just see who emerges as a winner, and then I'll buy them back. But they don't want to do the homework to find out for themselves who are going to be the winners. It's too easy to do it the other way.

**KYLE SHANNON**   There's a bit of a backlash. As it goes down, people are like, "Yeah, good—that'll keep you humble, you little pissants." My partner calls it revenge of the IT, but I think it's something deeper than that. I think there's a great resentment over the fact that this happened at all. So many people didn't understand what the Internet was, what these companies were doing, who all these snotty-nosed little fucks making millions are. On the West Coast you had a legacy of that kind of thing: people had seen young kids make their millions and get their Ferraris. Here, we haven't had that kind of entrepreneurial spirit since the turn of the century, since the industrialists. It's been forgotten. And because it's a media capital, what gets communicated out to the country is "Young whippersnapper dot-commers without a clue are making millions, you loser."

I think it's a temporary blip, but I think that if the Web doesn't stop sucking less, we could be in real trouble. I think the majority of the content you see on the Web is still painfully atrocious—too hard to navigate, too hard to get to, not compelling enough. If you need obscure research on some bizarre disease it's great, but if you want to choose between watching *Seinfeld* or surfing the Net, I'm going *Seinfeld*.

**JASON MCCABE CALACANIS**   We all expected it to come crashing down. Nobody expected it to come down as violently as it did. So that's all it is.

The world shifted. The rules before April 16 were: "Get big quick; be the first mover." So the rules after April 16 were, very clearly: "Become profitable; cut down your burn rates."

Each of those is actually unhealthy. "Take all the market share you can" is unrealistic. So you had companies trying to be all things to all people, and what happens is that you lose focus. And then you have today people saying, "Why, I only want to concentrate on my five top clients, and I don't want to do anything [risky]." Well, if you don't take any risk and chances, where is the growth? People are betting that these companies are going to grow, so if they are worth twenty times what their earnings are, their P/E ratio is twenty or thirty. You have to have some sort of miraculous growth, and if you are just focusing on the five or six that we make money off of today, then you'll never make it there. Both are irrational. And that's what is happening right now: irrational exuberance has led to irrational pessimism.

**MARC SINGER** The crash definitely hurts. It's going to make it harder to get money for newer companies and less proven companies. I'm sure if I was a VC I'd probably say "There are a lot of really bad companies out there, and this is a good thing." And there's some truth to that, but mostly I don't really believe that. I think that innovation takes a lot of risk. When you start a company, you passionately believe it's going to work. But no one knows one hundred percent that the product's absolutely going to work.

The cool thing about entrepreneurs is that they know what they can do, and they know it can be successful. But you're taking big risks, taking a leap of faith—"I'm going to do an advertising or a marketing product, and it will be really cool." But people told us we'd never be able to build the technology, we'd never get any clients, we'd never make any revenue.

There's a lot of skepticism, and that's great. But unfortunately, when you have a market crash like this, some of these people are going to have a much tougher time trying to build the space, to pitch a business that people aren't currently financing. Having said that, they're still better off in the Internet space than they are trying to build a new tennis shoe or something.

**ALAN MECKLER** A lot of these Internet companies aren't going to make it, and I think a lot of the young people in the Internet business are not money-savvy. See, I came up through the trenches, never had two nickels to rub together businesswise, so you always learn to wonder, "Is there going to be money there tomorrow?" Whereas many of the young Internet executives today really were clueless about the fact that there might not be money next year.

**ESTHER DYSON** I personally think it's good. The challenge is to discriminate between the good ones and the bad. For some people the correction will last forever, because they went out of business, and for others their stocks will be more fairly valued and they will go up over time. People are starting to be more rational, and what was going on was irrational.

**MARC SINGER** I think great innovation and new products often come out of nothing. It would be unfortunate if, because of the difficulty in raising capital now, someone just starting a great company can't do it.

There's also an age factor. In some ways you're better off if you've been in the business, have the connections, are a lot older (like forty years old), have saved up money, started a company. But on the flip side those people aren't going to start the next Netscape, they're not going to do what are for me the most interesting things, invent a new kind of entertainment. It sounds like "Don't trust anyone over thirty." But I just believe that.

So it's unfortunate if the climate becomes so severe that only those people can start companies. It's going to affect the kinds of companies that get started. That's my bias. But I'm not that pessimistic. I think it's still the Internet, it's still very early, and things go in a cycle. I don't think the April thing is just going to be forgotten. I think it's going to have lasting repercussions. But I do think people with great ideas will find a way to get them started.

**KYLE SHANNON** I quit acting to start this company. I want it to be here for a while. I want a company that I would want to come to work at, because I'll go back to acting if it starts getting boring. If it sucks, fuck that.

There were a lot of companies out there that were started just to make money, and if that's the only thing you're holding onto, then I think you're in trouble. We're holding on to this idea that we actually have the opportunity to deliver on the promise of the interactive future, whatever that might be. That's an awesome privilege and responsibility.

*One of the only major Alley successes in the year 2000 was the sale of About.com to Primedia for nearly seven hundred million dollars.*

**JERRY COLONNA**   If I was to tell you that I'm going to take a guy name Scott Kurnit, and after some hard attempts at Prodigy and MCI/NewsCorp [Delphi], he was going to go out, start a business, and two and a half years later sell it for seven hundred million dollars, you'd say, "Get the fuck out of here—that's not going to happen." But in fact that's exactly what the guy did. The problem is, in the intervening time the business was once worth two billion dollars. So our perception of failure is not based on what he accomplished—did he meet his goals and beat them?—but on this weird time frame, which everybody bought into. We bought into it on an individual basis; as I've said, *Time* magazine put Jeff Bezos on the cover as Man of the Year. There was a mediawide, almost a societywide, fascination with these kinds of personalities that was out of proportion. They were trying to do the hardest thing in business you can, which is start a new business.

*What will the environment be like in the future?*

**SETH GOLDSTEIN**   It will be really bad. Last April, when the market corrected, all the investors told their companies to make their money last until the end of the year. So now what we've seen is that a lot of companies that haven't been able to make it have exploded. There's another set of companies that are going to be given bridge loans to get through the winter. So January will be okay, and February will be okay, and March will see another group of companies explode because they can't last. April's going to be particularly bad. And then a long summer, and anyone who's left standing in the fall of 2001 will be a real company, and there's going to be very few of them—one out of ten.

**ALICE RODD O'ROURKE**   According to the NYNMA survey, in the greater New York area a quarter of a million [Internet] jobs have been created since 1996. There are a quarter of a million people who have worked in the industry, who are working in this industry, either full- or part-time, and it has built a base that we will continue to grow on. Yes, there have been companies that have laid people off, and that have closed, but we find that those people are finding positions quickly in other companies. They are needed and they don't want to go back to traditional industry.

**ESTHER DYSON** What we had was the age of the pseudo-entrepreneur, and everybody thinking he or she should run a company and become Bill Gates or Jeff Bezos. Now we're going to have the age of the real entrepreneur—people who build companies even though money is not being thrown at them, and who really want to build something rather than just be fashionable. Frankly, the world needs a lot of people besides entrepreneurs to work for the entrepreneurs. Not all people should start a business.

**SYL TANG** Right now, the social scene is completely, completely dead. People are feeling like they're swimming for their lives, so even people who are still in the industry are more focused on the work. They'd rather work now than party, because at any minute everything is going to blow up.

I have a shelf I call my "Dead Icon" or "Dot Bomb" shelf. It's full of relics I received from all the parties I went to. There's an iCast lunchbox, a Modo, a RocketBoard, and all these very expensive trinkets from parties I went to where they gave out these things for companies that are now dead.

I never left without a gift bag.

# HOW WE LOST THE WORLD

**B**y the closing days of the year 2000, the NASDAQ had lost more than half its value, and had been held underwater for eight full months—long enough for many companies to stop breathing. Across the Alley everything looked faded, aged, a step back, regressive. Many of the entrepreneurs picked their heads up from their work for the first time in half a decade, revealing sunken eyes and years of age, to find their empires in ruin, their entire Internet world up in flames.

The initial whirlwind of Silicon Alley may be over, and the beginning of its history closed. But where will the Alley go from here? For those who made some money, is there any chance to do it again? Or is Internet investment a one-trick pony that's had its day?

Those with the guts, vision, and wherewithal to make their millions in the early days deserve serious credit, though very few were able to cash out. Ambitious sorts like Craig Kanarick and Kyle Shannon may spend the rest of their lives trying to win back their paper fortunes—and who knows? They might succeed.

For those still struggling to stay afloat in these fickle economic waters,

*what needs to happen in order to survive? Is the excitement, the initial drive to
create and explore, gone for good? Has the Internet lived up to our expectations? Or did we expect too much from an industry in its infancy?*

**JASON CHERVOKAS**  I really miss the feeling of the early days.
The sense that we were remaking media, that it wasn't going to be the
old boys' network, which it turned out in fact to be. It wasn't supposed
to be mass media; it was going to be a media of the masses. The gatekeepers weren't going to be in control; it was going to be a democratic
way of having access to information. We were really experimenting,
and it was really, really exciting. I do miss that a lot.

**MARISA BOWE**  I never had any expectations. I didn't have a plan
for my life, I didn't have any particular dreams, I was just going into
whatever seemed interesting. I didn't have a dream of what the Internet should be. I had been through other media things before, and I
knew, as soon as I realized there was something to be commercialized,
that it was going to be commercialized alongside everything else,
because that's just how things work. With TV and music, the corporate
side of things sees a chance of making a profit off of something and
they seize it, wring the life out of it. But then out of some other corner
comes some fresh thing—people figure out a new way to make something cool and be heard.

**MARK STAHLMAN**  To the extent that there is a new media
industry actually to be built, I strongly suspect that New York will
have an important role to play in that. Precisely because it is Babylon.
And that's the part that has yet really to be seen. It will require a lot of
soul-searching for that to happen. And a lot of the folks here have
been too busy building whatever they were going to build for themselves to do much of that. Over the course of the next few years,
they'll have the chance to think a little bit harder and make some
longer-range plans.

**JOSH HARRIS**  I think the heyday was over two years ago.
There's no sense of group anymore. I feel like Old Timers' Day at Yankee Stadium.

**DAVID LIU**   A lot of bad ideas were funded and went away. I think it's part of the Darwinian process. I'm sure there were some funky-looking creatures roaming the Earth that were just never meant to survive, and this is the same kind of evolutionary process. People will say it was a time of great exuberance, of great optimism. I think the great retraction has occurred, but the little seedlings will pop up. The forest fire will actually replenish the space.

**RUFUS GRISCOM**   I think it's becoming increasingly possible to have an Internet job that's boring and uninspiring, and I think people will lose sight of this idea that you have a limited amount of time on this planet and there's something kind of exciting about being part of cultural change.

**ALAN MECKLER**   The legacy is very similar to other industry boom times we've seen. A lot of people don't realize that the birth of the telephone was a fairly exciting time, though it took longer to take root; the birth of the TV was a very exciting time; the birth of the automobile was a very exciting time. And the cities or places that it impacted—perhaps Detroit for the car, or whatever—probably had similar, if obviously smaller, numbers of people involved. And of course you didn't have the press or the media or the ability to get a story out in a second that we have today.

**JASON CHERVOKAS**   We went through a very distorted market period, where there was so much money available, funding bad ideas in a very herd-mentality kind of cycle, that no one got to test their ideas. If someone came out with an idea, twelve other guys were going to get funded who were going to compete with it. They were going to beat each other up. It's a lot like what happened with the U.S. and the Soviet Union in the '80s. Fight this war of attrition, bleed each other to death, and whoever runs out of money first loses.

**ALICE RODD O'ROURKE**   Do I think the investors blame the VCs, the investment bankers, and the entrepreneurs themselves? Well, I don't know. I blame only myself if I blame anybody. I for one assumed the rules of the universe could be bent for more than a brief moment. So

I don't know what other people are saying about it. The people I know who are involved in the industry feel, "I was richer; now I'm poorer." There is an excitement in knowing that you participated in this. As long as you didn't lose important money, I think people were accepting about the money they may have lost.

**CRAIG KANARICK** I think it's been incredibly educational. We have learned in six years what it took other industries thirty or forty years to learn. I actually think that it's really fortunate that we get all of this stuff out of the way early, so we can start operating like a real industry and get all this noise and bullshit out of the system—those people who thought they would get rich quick, or who just thought it was cool—and get the people who actually enjoy doing this stuff. It will stabilize soon.

**CLAY SHIRKY** One of my initial conditions of going to work at SiteSpecific—having been rattled by my experience at Columbia House—was that I'd never be asked to talk to anyone that didn't have an email address. You can't even imagine making that a condition of employment now, but at the time I just didn't want to talk to anyone who didn't get it. And that group, in the world of business, was an extremely small subset of people. And the email address was never from their business. It was always AOL or Panix or somebody. But now the Internet has completely pervaded all aspects of business life, so there almost doesn't need to be a Silicon Alley—because what Silicon Alley was about, by and large, was media-services business.

And, yeah, it's a little different from advertising, consulting, publishing—but not so different that it's off in this unimaginable sector of the economy. I don't want to get too elegiac; certainly there's more business than ever on the street. But I don't get the sense anymore . . . I'm as likely to be on the phone having a conversation with an A&R guy at EMI as with an Internet start-up.

We took the thing away from the people who didn't understand that it was their competition, built it up to the point where they suddenly understood it *was* their competition, and now, you know, they've bought in—literally or figuratively. In many cases they've acquired the assets of these companies. On some level the Web is going away the way the telephone has gone away: it's so ubiquitous you can't point to, "Oh, yeah, those businesses that use the telephone? I've heard of them. They're traded as a sector on Wall Street."

**SETH GOLDSTEIN** Everybody believed the hype. Everybody thought the rules had changed. We're all guilty. I guess there are some crotchety people who stayed away, and now are coming back saying "I told you so." Maybe *they're* not guilty.

**ESTHER DYSON** I thought the frenzy was stupid, disgusting, silly. But it probably was inevitable, and people are a lot wiser now. So I'm not terribly upset or distressed. It probably could've been done with a little less waste, and aesthetically I thought it was distasteful, but that's what the world is like. The only thing is, learning the lesson can be painful.

**KEVIN RYAN** I tell people—especially people who are really young—it's hard for them to fully appreciate it, but, you know, in a forty-year business career we won't go through a period like we went through in the last five years.

**SCOTT HEIFERMAN** I'm twenty-eight years old. This is not a time to reflect and be nostalgic. It's still the beginnings of everything. It could be one of these things where I'll realize later how unique and special it was. The most important answer would be that there was a lot of real innocence. I didn't close my eyes and think about a billion dollars, or the notion of what it would be like to have money.

**KYLE SHANNON** I miss our community's ability to talk about what was possible, rather than talking about quarterly numbers or turnover. When a lot of people get together and invent, amazing things happen. I miss the invention of that time. I know a lot of other cities have tried to re-create a WWWAC-like group, but I don't know that you can because the situation is so different. We know what the Web is now. You can go to school for it now. You can start a group and create a community, but I don't know that you can start a community of people who literally *all* said, "We don't have a freaking clue, but let's go do it anyway."

I'm glad we're not in it anymore, because it did grow up. There was a lot of uncertainty, a lot of pain; a lot of mistakes were made back then. We're starting to figure it out. But one of the things we need to keep asking ourselves is, "Is it good enough? Can we keep inventing?"

**JEREMY HAFT**   Here's an example of Internet culture on business culture in general: the pervasion of business casual into the world of dress. That was an Internet innovation, and now you find law firms and accountancies and everyone's trying to imitate Internet guys— which I would imagine came about from technologists who came in with ripped and torn and mismatched clothes, and that's sort of become the dress code.

*How many people in Silicon Alley made money?*

**SETH GOLDSTEIN**   A fair number. It depends what you consider real money. A lot of people were gainfully employed. A lot of people got paid really well, even at an hourly rate: consultants were able to earn fifteen hundred dollars a day when really they were worth five hundred, or made three hundred thousand dollars a year when they really deserved one hundred thousand dollars. There was a lot of that across the board. There were far fewer people who actually pocketed millions of dollars, who sold something for cash, sold something for stock, sold the stock. I think there's a much smaller percentage of Silicon Alley that was fortunate enough not to leave much money on the table. A core few did disproportionately well, but a lot of that money is going to run out, depending on how fast people spend it in the next couple of years. So what will be interesting is, what will some of these entrepreneurs do next?

**MARISA BOWE**   I don't feel disappointed, I feel like Robin Hood. I got to do this for five and a half years. I got to be paid good money and have this relatively big budget to do it. And I knew what was happening, that this was a really rare opportunity, and so I wanted to capitalize on it by doing it creatively.

**DAVID LIU** The one thing I am disturbed by—and I hope this isn't our legacy—is that innovation, new products, and new businesses will become the domain of large conglomerate corporations. One of the things we were really excited about with the Internet was that *People* magazine took seven years to break even; *Sports Illustrated* took ten. TimeWarner could afford to do it. They knew that over time the brand would gain some value. But suddenly investing in, or becoming part of, something in its infancy became attractive to the consumer.

**DOUGLAS RUSHKOFF** I think we very innocently stumbled into the darkness of business. We didn't know. We didn't realize what businesses can do, and what they were. All we realized was that adults didn't believe that we had done something, we had stumbled upon something terrific. And getting AT&T to come online and build a website was great confirmation: "See, this is real, now they're coming into our world." What we didn't realize was that they were bringing us into their world.

**RUFUS GRISCOM** I think in '97 I felt like, Wow, this has already played itself out—there's no mystery here. But there *was*, back in those early days; it was challenging just to survive as an Internet company. The real difference is that there weren't all these capital infusions. The idea of raising millions of dollars in 1997 never even crossed our minds. I mean, millions of dollars? Who on earth would give us millions of dollars? The idea that anyone would give a college grad millions of dollars was unthinkable at the time, and that in some ways lowered the stakes. Because there were fewer powerful financial incentives and exigencies, there was more freedom to do what you wanted, and a little bit less pressure. The pressure was just to survive at all.

There was this sense that you were crawling in a desert, dragging your canteen, and you'd run into someone else crawling in the desert and be delighted to see them and you'd go have drinks together. There wasn't a sense that you had competition. Word and FEED might have been perceived as our competition, but we didn't think of it that way at all. We were sort of delighted to have company that was continuing to exist and live and breathe with us. It provided evidence that *we* would continue to live and breathe and survive. Continuing with this metaphor of western expansion and of being a pioneer, you reach a point where all of a sudden people are building fences and digging in their territory and keeping people off their property—which, you know, is sort of logical, and also means there's economic value to your property. And that's exciting.

**STEFANIE SYMAN**  Having lived through the early days of Silicon Alley, it really was a kind of unrepeatable moment watching the birth of a medium, however flawed it was. Now we're actually in its adolescence. A lot of companies have died; everyone's now pissed off at it and it's pissed off with itself. A lot of angst and will-this-go-anywhere concerns. It's like if you're city kids camping—you're doing something that's a little bit beyond your actual ability, and yet because you go through it with other people fumbling around with you, you end up having a connection to the people and the place that goes beyond just having a job with someone or living in New York at the same time.

**DAVID LIU**  We should not be surprised that people lost a lot of money, because guess what? A lot of people *made* a lot of money. It's all evened out. What I am afraid of is if the market doesn't recover—if we're dragged into an oil crisis or something. Then we're back to the big companies being the ones who are able to invest and lose money before they can build something up. I would hate for that to be the legacy of the last five years.

**JASON CHERVOKAS**  Clearly the network of networks is here to stay. TCP/IP is the revolution, not any one part of it. And this new many-to-many way of communicating is here to stay. We've only begun to experiment now with ways of using that. And what we see happening with Napster and peer-to-peer is a further move out on the path of concentric circles from this Big Bang of opening instantaneous global communications without gatekeepers. I don't think we've really begun to say, "Okay, this is the new world we're living in, where messaging is the medium, and every person and every point on this network of networks is both client and server." The implications of that for the way we relate to each other interpersonally, and for the way we relate to information as we once thought of it—we haven't really begun to build around those new assumptions yet. Silicon Alley jumped into this with both feet as a communications platform; that was the explosion.

**JACK HIDARY**   The whole industry's a roller coaster. If you don't like roller coasters, get off. I didn't like roller coasters as a kid, but I guess I got to like them. I tell you, you never get bored. We have created in this industry, both here in New York and elsewhere, a cadre, a group of young leaders, managers, that normally would have taken twenty years to get up the chain of management programs and training programs and God knows what else. And we just put them in the fire. We threw everything at them—public offerings, hiring people and firing people, all these kinds of things. We've thrown a complete business training at them in two or three years.

A lot of people pooh-pooh certain companies—those that went out of business—but great learning has happened. And yes, a lot of companies should not have been public, and yes, a lot of companies should never have gotten funded. But that's part of an initial cycle: You spread a lot of seeds and you see what grows. That's part of the ball game. In 1981 there were three thousand software companies, and today there are five. And that's just the way it is.

**ALICE RODD O'ROURKE**   As bright and ambition-filled as the people in the new media are, they have not been able to change the rules of the universe. And the rules of business are among them. One of those is that a certain of amount of investment has to yield a certain amount of return within a certain amount of time, and that has typically been three to five years, maybe a few more. And that's a rule of business that we're not able to bend for longer than a brief period of time. But as I said, this was a very important period for the industry. After all, we have a part of town named after us. Silicon Alley is no longer a location, it's a state of mind.

**JOSH HARRIS**   I think we did it. We did it and . . . you know, it just changed. And I'm just going through different people I know. Some people are still in the game. Look at Marc Bell at Globix. He's in the game; he wants to grow his company. And then I look at other people who I won't name, who are still working, who I know have their exit strategy worked out. They made their money, now they're thinking, "Why am I here? Why do I need to keep doing the same thing over and over again?"

**ANDREW RASIEJ**  I think people are starting to burn out. It's cyclical. People who have been doing it for some time are starting to say, "There has to be more than this." And then you've got a lot of young people who are involved in this; now those people are reaching their late twenties or thirties, and they're thinking about getting married, and their spouses aren't necessarily in new media and are making them compromise, choosing between their personal lives or business lives. I don't know how much of it is burnout on this particular type of technology, or whether it's because people are getting older. I don't think there are any seventeen- or eighteen-year-olds who are networked to each other experiencing burnout. Although there are factions of Luddites out there, kids who are saying "We don't want to have anything to do with this, because there's just too much information." They're looking at an Amish model. I was part of a generation that was somewhat liberal and progressive. Our children, or my peers' children, are probably more in line with our grandparents—are probably more puritanical, conservative. It's cyclical.

**MARISA BOWE**  It's like an ant farm—watching everyone going off in these different directions, and what is fascinating is seeing the unfolding of character over time.

Someone like Jason Calacanis—What would he have done? He needed something as big as this to develop himself to the full extent that he did. And the same goes for me, in a way. I have the same brains as before, but I was typing away in Echo, performing for that tiny little world, with no chance to do a book or a game or any of that stuff.

**RUFUS GRISCOM**  I would say we're experiencing the beginning of the Internet revolution right now. This is just the beginning. Fifteen years from now we'll look back to when we were sitting around this table and think, "Wow, it really was the beginning."

**CLAY SHIRKY** That kind of crazy, flexible, who-knew-what-would-happen period is over, and it's never coming back. I've spent the last year writing and thinking about Napster, and for a lot of us Napster has reenergized us. But on some level it feels like the drunk old gunfighter strapping on his six-guns one last time. Napster really feels like, "Ah, someone has really come along and fucked something up." I remember when the Internet was like that every day. Now mostly you get up and the Internet's doing things like making it cheaper to buy pet food. And that's a thing that needs to happen, but it isn't like the old days.

**SCOTT KURNIT** If you look at that orange building over there [points out his window], that is the former Wang building. And if you look, you can also see the Met Life symbol on what was once the Pan Am building. So from this very spot I am constantly reminded that as big and as bold as you think you get, you can get crushed in a moment. It keeps you appropriately paranoid, and it's critical, critical, to have that kind of history. Wang owned the desktop in America if not the world at one time, and now doesn't exist. I remember walking past it as they were scraping the Wang logo off.

**MARK STAHLMAN** What I decided was we were going through an almost agricultural turning over of the soil—going through a series of harvests, building companies and watching them fail, building up the capacity to meet payroll and then get into real difficulties, being able to generate some real wealth. And then discovering who among the people able to accumulate that wealth were satisfied with it, who just wanted to get their picture on the front of a magazine, who just wanted to get the supermodels, who just wanted to buy the Ferrari, and who was actually in it for the longer run. Because until people are upset with existing media, then they can't build a new media industry.

**MARISA BOWE** I don't think it's over yet. I think it's sort of like Balzac's novel *Lost Illusions*. Don't take the title too seriously, but it's this incredible tableau of Parisian society during this Gilded Age, seen through the eyes of this provincial kid who wins a poetry contest and so gets involved with the big city society. And there's a lot of people manipulating the stock market and journalism in order to get where they want to go. Mass journalism was new then, so it was an opportunity for people who didn't have any money to get somewhere if they were smart. And this was a huge development. All these people realizing that they could get somewhere, that here were a bunch of rides being offered to them.

**MARK STAHLMAN**   There have been those who had an interest in building a new industry. And there have been those who couldn't care less about any of that, but for whom fame and fortune is the principal objective. A lot of work has been done on entrepreneurs over the years, and I think the research on entrepreneurs is pretty uniform. People don't do it to get rich and famous. The successful ones are really thinking about sweeping new technologies and broad-scale changes.

**CLAY SHIRKY**   If anything is going to characterize this particular moment in time, it's that the people whose lives were changed by it have adopted the Internet as a kind of second language. We know more about it or think more natively in Internet than the people we have followed, but less natively than the people who are following us. And we'll probably spend a good deal of the rest of our lives translating from the younger generation to the older generation, until the older generation has also adopted this second language, or the population changes over, and it becomes like the automobile or the phone—just such a part of normal American life that no one can imagine it ever didn't exist.

**SCOTT HEIFERMAN**   I have three brothers and a sister, all in their forties. None of them were online three years ago, and now they all are. And it's changed their lives. One of my brothers has three kids, and they don't have a CD player in their house. When my nine-year-old nephew Jeffrey thinks of music, he doesn't think in the form of a CD or album or cassette. He thinks of music as a file on your computer, and you download it. Forty years from now his brain will be wired in a different way, because of that.

**JOHN YOUNG**   When you look at the amount of talent and money it takes to produce a thirty-second television spot—all the millions in an edit suite to put it together, then multimillion dollars to upload it through all these satellite systems, broadcast it out to this huge network of things . . . But we *don't* think about that. We just think about what we see on TV. And where the Internet needs to go is to stop thinking about, Oh, it's a database or HTML or DSL or whatever—it's about what's delivered.

**NICK NYHAN** There are certain people out there who are enthusiasts for anything that's new. But there's a growing number of people who don't want new, they just want the Web to work efficiently. They love going to read their newspaper and seeing the Red Sox score. They don't want a million bells and whistles; they don't want to be overmarketed. In order for the Internet to grow up it has to learn. Just like when you're a teenager and you're trying to make yourself cool, there's a certain point where you stop and say, "I am who I am, and I'm not going to pierce my nose just to be cool."

*What will be the legacy of Silicon Alley?*

**JERRY COLONNA** Lovey, from Pseudo.com [laughs uncontrollably].

**FRED WILSON** The first new industry created in New York in the past fifty years.

**JERRY COLONNA** The fact is that all of the people who are business leaders here are going to be business leaders in New York for the next two or three decades. And that will be the long-term legacy.

**ALAN MECKLER** I look at the Internet as a giant tiger. And I was—and I'm still—holding on to the tail for dear life. And it's pushing me and dragging me and throwing me all over the place, but I'm holding on, and I'm making money. There are those who also grabbed on to the tail initially, but started to believe they were the tiger. And they're the ones who are now in trouble. But I still think the Internet the most exciting industry in the history of the world.

# GLOSSARY

**Angel investor.** An individual investor who invests in very-early-stage companies, generally at the idea stage or within the first few months of its life. Angel investments are highly speculative and can often come at a high premium to the entrepreneur, but are invaluable in building a company to prepare it for an institutional investment.

**Book.** A tally or roster of nonbinding reservations to purchase shares in an IPO. These reservations are collected by financial institutions during the period leading up to the IPO, when entrepreneurs go on road shows in search of investment capital.

**Bootstrap.** To grow a company organically without significant outside funding. Alan Meckler called bootstrapping one of the key secrets to successful Internet companies.

**B2B.** Business to business. Term used to describe companies that cater to business customers instead of consumers, and most specifically companies that create virtual online marketplaces for established industries, such as eSteel for the steel industry, or VerticalNet for several industries.

**B2C.** Business to consumer. Invented to differentiate B2B sites from B2C, and thus often used disparagingly.

**Bandwidth.** The amount of information that can be transferred in a given amount of time. The more bandwidth one has, the faster one can download or transfer files.

**Bricks and Mortar** or **Clicks and Bricks.** An integrated online/offline business, like Toysrus.com, or BN.com.

**Broadband.** High-speed access to the Internet.

**Box.** A computer. A "Linux Box" is a computer using a Linux-based operating system.

**Burn rate.** A company's monthly total expenses. The burn rate can be used to calculate the amount of time a company has left before it goes under, a formula all dot-com entrepreneurs know well: remaining funds divided by burn rate equals remaining time (in months) until bankruptcy.

**Cannibalize.** Generally, to publish a company's content online for free when it's being sold offline at a cost—an issue that caused dilemmas at all of the major media companies in the face of the Internet opportunity, as media companies like Time Warner, Condé Nast, Disney, and others struggled for years to make sense of their Internet strategies.

**Click-through.** To click on a banner ad; the total number of clicks on the ad. Click-through was a double-edged sword for supporters of online advertising. While it provided, for the first time, a method of direct response to an advertising message, declining click-through rates made online ads seem worthless. Click-through rates fell to less than half of one percent by 2000 for any given banner ad.

**Click-stream.** The sequence of choices (clicks) that a user makes on any given site, which can be collected as data and analyzed.

**CD-ROM.** Compact Disc Read-Only Mechanism. The CD-ROM industry played a major role in early Silicon Alley, but was displaced by the Web industry—which solved the nagging problem of distribution by using the Internet instead of a packaged disk on a store shelf to get to a user's personal computer.

**CGI.** Common Gateway Interface.

**CPM.** Cost per thousand. A twenty-dollar CPM for banner ads would cost the advertiser twenty dollars for every thousand banner ads shown on a website. One of the biggest challenges of the content companies was the failure of banner advertising to generate enough revenue to support costs. CPM rates dropped steadily.

**Dark.** Down. When a website goes dark, it is no longer accessible from the Internet.

**Due diligence.** A close look at a company's financial health and the background of its officers, generally performed by an institutional investor before an investment is made.

**E-commerce.** Electronic sales, generally to consumers. Though it was once the darling of Wall Street, when Amazon.com spawned a horde of imitators for every consumer good (pets, clothing, and CDs were especially disastrous), it was one of the first sectors to take a dive in the late nineties, helping to tip off the market crash.

**Exit strategy.** Plan to cash out for investors and shareholders: "Our exit strategy is to be acquired for four billion dollars by Yahoo."

**FTP.** File Transfer Protocol. Used for sending files across the Internet.

**HTML.** HyperText Markup Language. The design and publishing metalanguage of the World Wide Web, HTML is a fairly simple set of codes designers use to make the Web into a visual environment. Developed from SGML, a textbook-publishing metalanguage.

**HTTP.** HyperText Transfer Protocol. The main protocol of the World Wide Web, which describes how messages should be sent, and how a user and a website interact on the technological end.

**HyperText.** A system for linking objects to each other—for example, a phrase in a sentence on a Web page that, when clicked, delivers the user to a definition.

**Insider selling.** The selling of shares of stock by founders, early investors, and any other director of a public company.

**IPO.** Initial Public Offering.

**ISP.** Internet Service Provider.

**Java.** A programming language well suited for the Web.

**Market cap.** What a public company is worth based on the total value of its shares.

**Media play.** An online media company. See *play*.

**New media.** Electronic, digital, and online art, content, and businesses.

**Options.** The right to buy stock at a fixed price.

**Packet switching.** One of the fundamental concepts of the Internet, this is the method of moving information across a network by breaking it into small pieces (packets), and reassembling them on the recipient end.

**Page views.** The total number of Web pages accessed by users at a website—not to be confused with "hits," which counts every part of the page (graphics, etc.) individually.

**Paper millionaires.** Entrepreneurs who have millions of dollars' worth of options and shares in their company, but not necessarily any cash.

**Permission Marketing.** When consumers give permission to marketers to send them targeted information.

**Play.** The basic strategy or business method of a company, often used in describing the viability of a business model: "We're considering investing in a bicycle-messenger play."

**Pure play.** A company whose sole focus is the Internet, such as an online-only bookstore. When Wall Street and VC valuations for companies were still in flux for the dot-coms, to be a pure play was the ultimate valuation maker; after the market crashed, it spelled disaster.

**Repurpose.** To reuse already published content on the Web, or in another format. See *shovelware.*

**Road show.** A two-week period in which entrepreneurs tour financial institutions around the country to fill investment in its IPO. Generally performed in a private jet hired by the underwriter, with three or more stops and presentations per day.

**S-1.** Initial Registration Documents. Documents prepared with an underwriter in preparation for an IPO, these are provided to investors and the Securities and Exchange Commission (SEC).

**SEC.** Securities and Exchange Commission. Responsible for governing the public markets, including all IPOs.

**Server.** Computer that makes website information available to the Internet.

**Shovelware.** Content from an existing medium used in another medium, or software thrown in to take up additional space on a CD-ROM or package.

**Spam.** Junk email.

**Stickiness.** The amount of time spent by a user on a website. The stickier the site, the more time users spend on it.

**Tables.** Introduced into Netscape 1.1 in 1995, tables allowed Web designers to align and place text and images on Web pages more precisely through the use of rows and columns.

**TCP/IP.** Transfer Control Protocol/Internet Protocol. The set of protocols that describe how information travels over the Internet. The standard for sending data from one computer to another, TCP/IP is a part of the UNIX code.

**Term sheet.** An offer to purchase shares, given by a potential investor to a start-up, the term sheet describes the terms of the transaction—setting the valuation and stipulating how many seats on the board of directors will be given to the investor.

**Quiet period.** The time between a company's filing its S-1 document and the day it goes public, during which it is not allowed to make forward-looking public statements. Since entrepreneurs were forced to stay out of the press, they spent heavily on television ads during this period.

**Underwriter.** Investment bank that takes a company public.

**UNIX.** The leading operating system for workstation computers (for engineering and programming), developed at Bell Labs in the 1970s.

**Usenet.** An Internet-based bulletin board system with newsgroups that cover thousands of topics.

**Valuation.** The perceived value of a company, based on the number of shares times their value at the most recent trade. For a pre-IPO company, valuation is set by the latest round of investment. For example, when Flatiron Partners first invested three million dollars in StarMedia, it was at a nine-million-dollar valuation.

**Vaporware.** A software product that is marketed and hyped in the press so far in advance of release that many suspect it may never come to pass.

**Vesting.** Period of time before an employee's options become available. Many employees in start-ups vest quarterly over a four-year period: as long as they stay with the company, then, they can exercise another quarter of their shares for each of the first four years.

**VC.** Venture Capitalist. Risk investor who buys shares of private companies to help build them up and prepare for an IPO or sale.

**WWW. World Wide Web.** The visual and multimedia aspects of the Internet, viewed through a Web browser (like Netscape or Internet Explorer), serving HTML-based pages. Though the word "Web" is often used interchangeably with "Internet," it only came into existence in 1991 with the invention of Tim Berners-Lee's archetypal browser at CERN, the European Nuclear Physics Laboratory, in Geneva.

# ACKNOWLEDGMENTS

The interviews that compose the text of *Digital Hustlers* were made during the summer, fall, and winter of 2000–2001. We would like to thank everyone we interviewed and spoke with while working on the book. The entrepreneurs, employees, journalists, investors, artists, and others who devoted their time and memory to telling the history of Silicon Alley did so with a great deal of trust and openness.

Special thanks to Jim Fitzgerald for being the first to encourage this project and for being a friend throughout. With gratitude to Calvert Morgan for his belief in the work and his patience, and to everyone at ReganBooks and the Carol Mann Agency for their efforts.

Philip Angell, CEO of HearingRoom.com, used his real-time voice transcription technology to transcribe several of the interviews, and we would like to thank him for his generosity and interest.

We would also like to thank Caleb Furnas and Suzy Hansen for their important contributions. For their support, Dana Bilsky, Jon and Sam Effron, Jessica Blank, Nicholas Hallett, William O'Shea, Ewald Christians, Jeremy Saland, Aaron Matz, Gary Hustwit, Gabriel Snyder, Kevin Spett, Janey Choi, Toure Folkes, and always Sumi Abeysekera.

In loving memory of Morris Arfa, George and Mary Kait, Elizabeth Pupain, and Andor Weiss, and with gratitude to our families, especially Daniel Weiss, Gertrude Arfa, and Sylvia Weiss.

# INDEX